Reprint Publishing

For People Who Go For Originals.

www.reprintpublishing.com

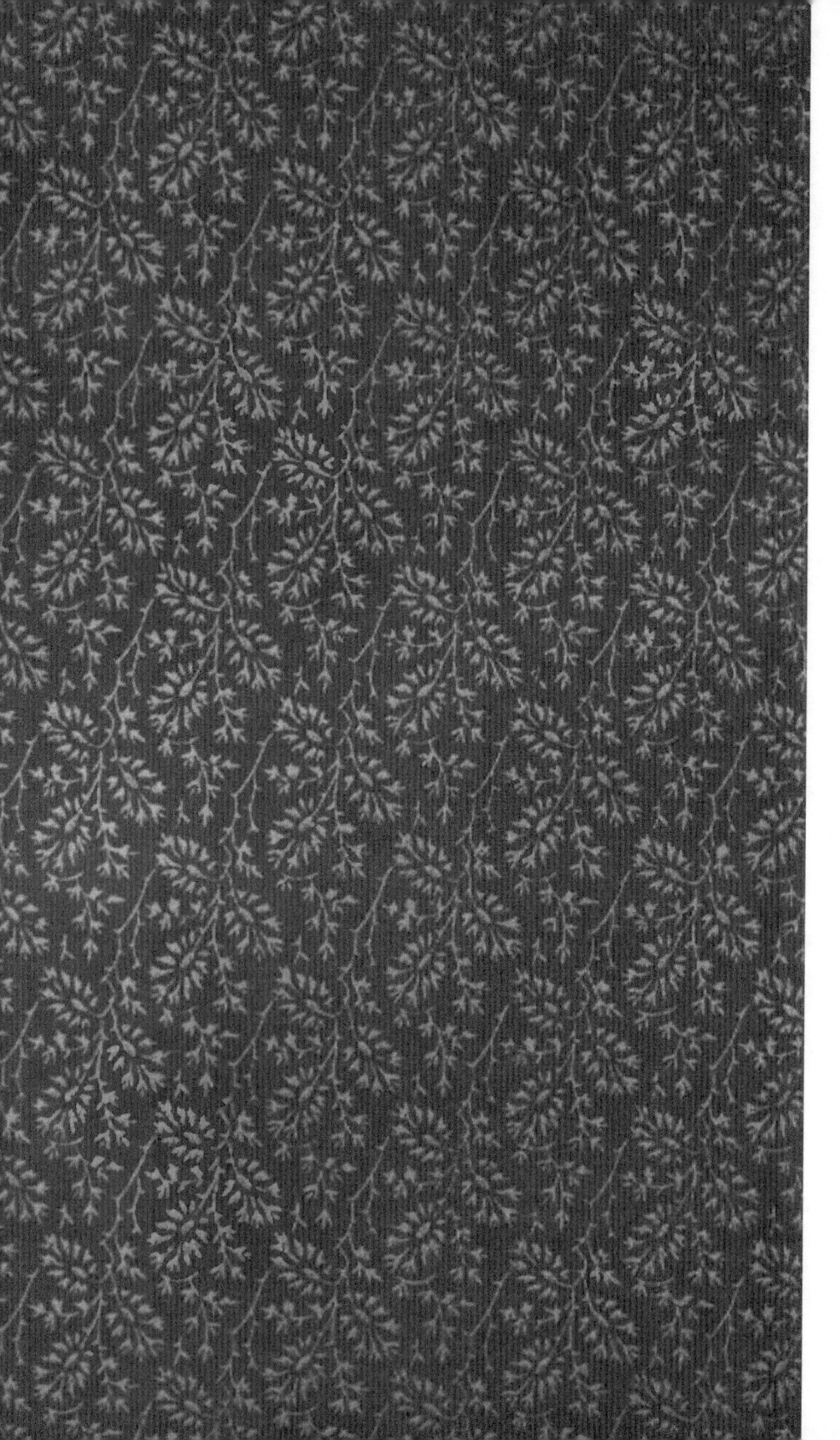

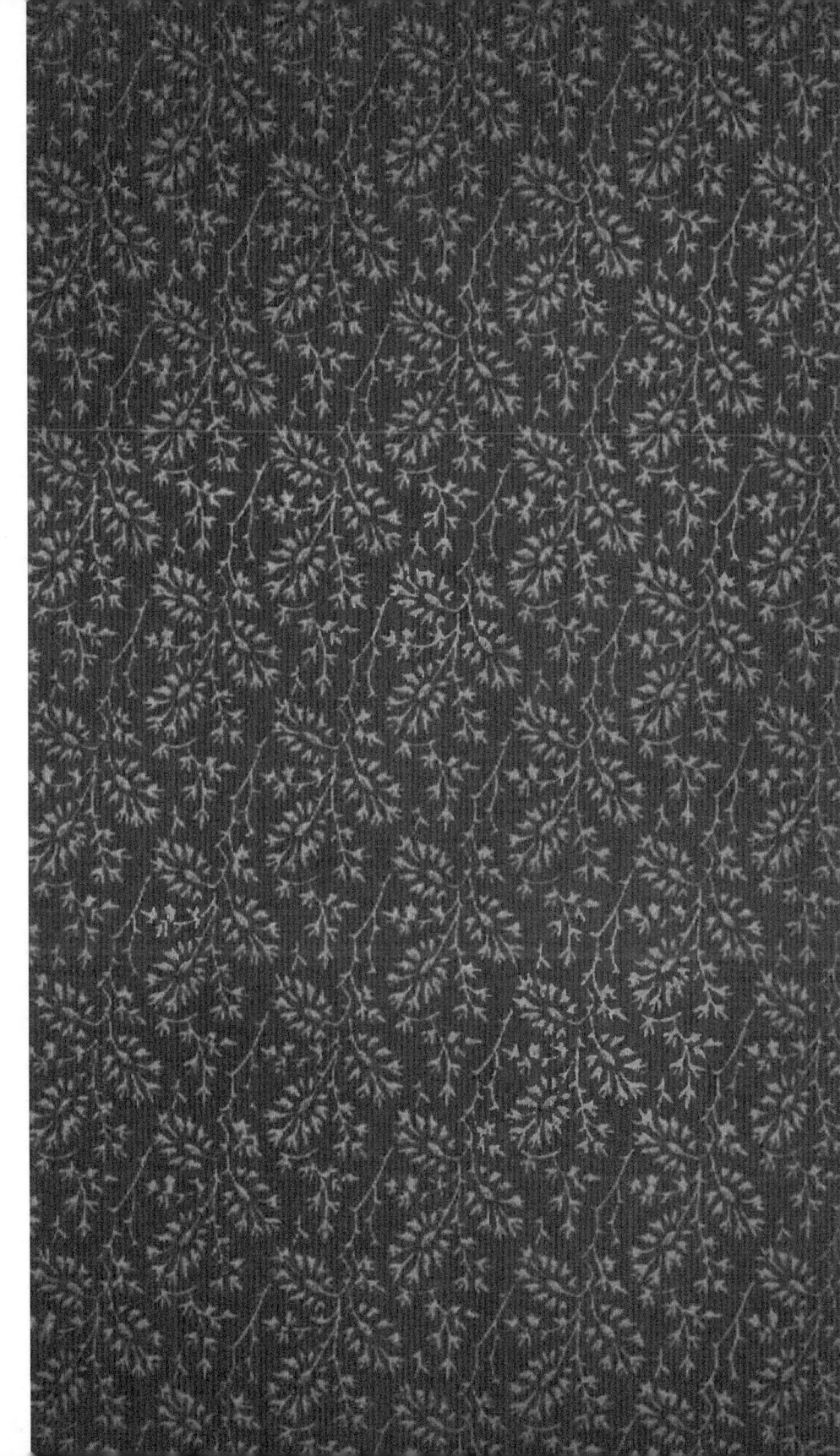

"The wicked borroweth and returneth not again."

PUBLISHER'S NOTE.

Two hundred and ten copies of this Work printed on superfine Royal 8vo paper.   Each copy numbered.   Type distributed.

No.

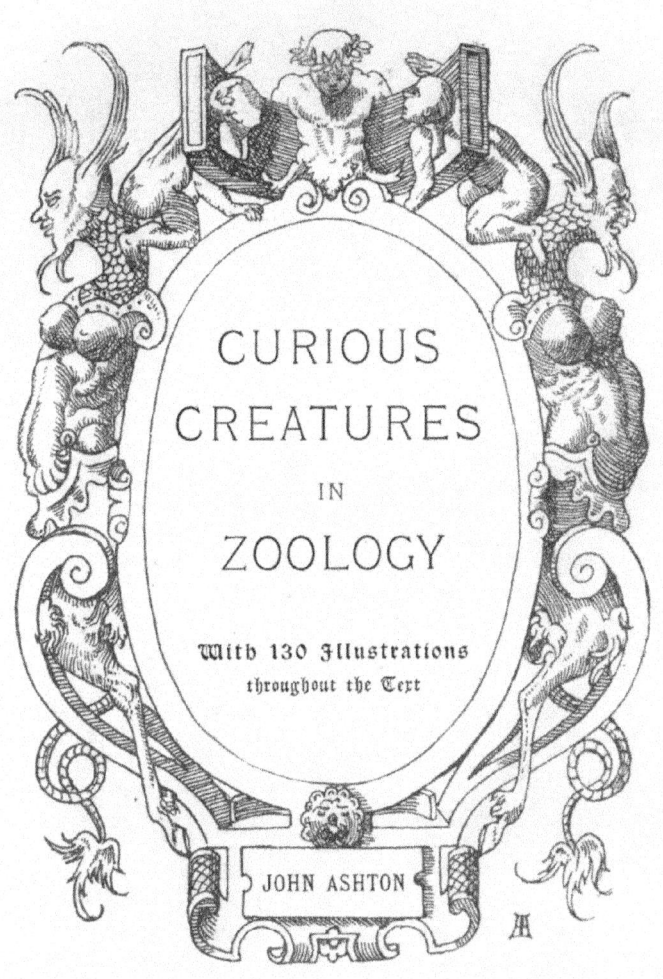

CURIOUS CREATURES
IN
ZOOLOGY

With 130 Illustrations throughout the Text

JOHN ASHTON

LONDON
JOHN C. NIMMO
14, KING WILLIAM STREET, STRAND
1890

PREFACE.

"Travellers see strange things," more especially when their writing about, or delineation of, them is not put under the microscope of modern scientific examination. Our ancestors were content with what was given them, and being, as a rule, a stay-at-home race, they could not confute the stories they read in books. That age of faith must have had its comforts, for no man could deny the truth of what he was told. But now that modern travel has subdued the globe, and inquisitive strangers have poked their noses into every portion of the world, "the old order changeth, giving place to new," and, gradually, the old stories are forgotten.

It is to rescue some of them from the oblivion into which they were fast falling, that I have written, or compiled, this book. I say compiled it, for I am fonder of letting old authors tell their stories in their old-fashioned language, than to paraphrase it, and usurp the credit of their writings, as is too much the mode now-a-days.

It is not given to every one to be able to consult the old Naturalists; and, besides, most of them are written in Latin, and to read them through is partly unprofitable work, as they copy so largely one from another. But, for the general reader, selections can be made, and, if assisted by accurate reproductions of the very quaint wood engravings, a book may be produced which, I venture to think, will not prove tiring, even to a superficial reader.

Perhaps the greatest wonders of the creation, and the strangest forms of being, have been met with in the sea; and as people who only occasionally saw them were not draughtsmen, but had to describe the monsters they had seen on their return to land, their effigies came to be exceedingly marvellous, and unlike the originals. The Northern Ocean, especially, was their abode, and, among the Northern nations, tales of Kraken, Sea-Serpents, Whirlpools, Mermen, &c., &c., lingered long after they were received with doubt by other nations; but perhaps the most credulous times were the fourteenth and fifteenth centuries, when no travellers' tales seem too gross for belief, as can well be seen in the extreme popularity, throughout all Europe, of the "Voyages and Travels of Sir John Maundeville," who, though he may be a myth, and his so-called writings a compilation, yet that compilation represented the sum of knowledge, both of Geography, and Natural History, of countries not European, that was attainable in the first half of the fourteenth century.

All the old Naturalists copied from one another, and

thus compiled their writings. Pliny took from Aristotle, others quote Pliny, and so on; but it was reserved for the age of printing to render their writings available to the many, as well as to represent the creatures they describe by pictures ("the books of the unlearned"), which add so much piquancy to the text.

Mine is not a learned disquisition. It is simply a collection of zoological curiosities, put together to suit the popular taste of to-day, and as such only should it be critically judged.

JOHN ASHTON.

CONTENTS.

	PAGE		PAGE
INTRODUCTORY	1	THE GORGON	83
AMAZONS	23	THE UNICORN	87
PYGMIES	26	THE RHINOCEROS	97
GIANTS	32	THE GULO	101
EARLY MEN	38	THE BEAR	105
WILD MEN	44	THE FOX	125
HAIRY MEN	47	THE WOLF	134
THE OURAN OUTAN	51	WERE-WOLVES	140
SATYRS	55	THE ANTELOPE	145
THE SPHYNX	61	THE HORSE	146
APES	65	THE MIMICK DOG	150
ANIMAL LORE	67	THE CAT	154
THE MANTICORA	71	THE LION	156
THE LAMIA	74	THE LEONTOPHONUS—	
THE CENTAUR	78	PEGASUS—CROCOTTA	157

CONTENTS.

	PAGE		PAGE
THE LEUCROCOTTA—THE EALE—CATTLE FEEDING BACKWARDS	159	TWO-HEADED WILD GEESE	203
		FOUR-FOOTED DUCK	203
ANIMAL MEDICINE	160	FISH	206
THE SU	163	MERMEN	206
THE LAMB-TREE	165	WHALES	214
THE CHIMÆRA	170	THE SEA-MOUSE	234
THE HARPY AND SIREN	171	THE SEA-HARE	234
THE BARNACLE GOOSE	174	THE SEA-PIG	235
REMARKABLE EGG	179	THE WALRUS	235
MOON WOMAN	180	THE ZIPHIUS	238
THE GRIFFIN	180	THE SAW FISH	239
THE PHŒNIX	183	THE ORCA	239
THE SWALLOW	186	THE DOLPHIN	242
THE MARTLET, AND FOOTLESS BIRDS	189	THE NARWHAL	244
		THE SWAMFISCK	245
SNOW BIRDS	191	THE SAHAB	247
THE SWAN	193	THE CIRCHOS	247
THE ALLE, ALLE	194	THE REMORA	253
THE HOOPOE AND LAPWING	196	THE DOG-FISH AND RAY	255
		THE SEA DRAGON	256
THE OSTRICH	197	THE STING RAY	256
THE HALCYON	199	SENSES OF FISHES	258
THE PELICAN	200	ZOOPHYTES	259
THE TROCHILUS	201	SPONGES	260
WOOLLY HENS	202	THE KRAKEN	261

CONTENTS. xi

	PAGE		PAGE
CRAYFISH AND CRABS	267	THE SALAMANDER	323
THE SEA-SERPENT	268	THE TOAD	326
SERPENTS	278	THE LEECH	329
WORMES AND DRAGONS	293	THE SCORPION	330
THE CROCODILE	311	THE ANT	332
THE BASILISK AND COCKATRICE	317	THE BEE	332
		THE HORNET	333

INDEX 335

CURIOUS CREATURES.

LET us commence our researches into curious Zoology with the noblest of created beings, Man; and, if we may believe Darwin, he must have gone through many phases, and gradual mutations, before he arrived at his present proud position of Master and Conqueror of the World.

This philosopher does not assign a high place in the animal creation to proud man's protogenitor, and we ought almost to feel thankful to him for not going further back. He begins with man as an Ascidian, which is the lowest form of anything of a vertebrate character, with which we are acquainted; and he says thus, in his "Descent of Man":—

"The most ancient progenitors in the kingdom of the Vertebrata, at which we are able to obtain an obscure glance, apparently consisted of a group of marine animals, resembling the larvæ of existing Ascidians. These animals probably gave rise to a group of fishes, as lowly organised as the lancelet; and from these the Ganoids, and other fishes like the Lepidosiren, must have been developed. From such fish a very small advance would carry us on to the amphibians. We see that birds and reptiles were once intimately connected together; and the Monotremata now, in a slight degree, connect mammals with reptiles. But no one can, at present, say by what line of descent the three higher, and related classes—namely, mammals, birds, and reptiles, were derived from either of the two lower vertebrate classes, namely, amphibians, and fishes. In the class of mammals the steps are not difficult to conceive which led from the ancient Monotremata to the ancient Marsupials; and from these to the early progenitors of the placental mammals. We may thus ascend to the Lemuridæ; and the interval is not wide from these to the Simiadæ. The Simiadæ then branched off into two great stems, the New World, and Old World monkeys; and from the latter, at a remote period, Man, the wonder and glory of the Universe, proceeded."

"We have thus far endeavoured rudely to trace the genealogy of the Vertebrata, by the aid of their mutual affinities. We will now look to man as he exists; and we shall, I think, be able partially to restore during successive periods, but not in order of time, the structure of our early progenitors. This can be effected by means of the rudiments which man still retains, by the characters which occasionally make their appearance in him through reversion, and by the aid of morphology and embryology. The various facts to which I shall here allude, have been given in the previous chapters. The early progenitors of man were no doubt once covered with hair, both sexes having beards; their ears were pointed and capable of movement; and their bodies were provided with a tail, having the proper muscles. Their limbs and bodies were also acted on by many muscles, which now only occasionally reappear, but are normally present in the Quadrumana. . . . The foot, judging from the great toe in the fœtus, was then prehensile; and our progenitors, no doubt, were arboreal in their habits, frequenting some warm, forest-clad land. The males were provided with great canine teeth, which served them as formidable weapons."

In fact, as Mortimer Collins satirically, yet amusingly, wrote :—

> "There was an APE, in the days that were earlier;
> Centuries passed, and his hair became curlier,
> Centuries more gave a thumb to his wrist,—
> Then he was MAN, and a POSITIVIST."

The accompanying illustration, which seems to embody

all the requirements of Darwin, as representing our maternal progenitor, is from an old book by Joannes Zahn, published in 1696—and there figures as "Ourani Outains."

Darwin says that the men of the period wore tails,

and if they were no longer than that in this illustration (which is copied from the same book), they can hardly be said to be unbecoming — still that is a matter for taste—they are certainly more graceful than if they had been rat-like, or like a greyhound, or toy terrier. Many old authors speak of tailed men in Borneo and Java, and not only were men so adorned, but women. Peter Martyr says that in a region called Inzaganin, there is a tailed race—these laboured under the difficulty of being unable to move them like animals —but as he observes, they were stiff like those of fishes and crocodiles—so much so, that when they wanted to sit down, they had to use seats with holes in them.

Ptolemy and Ctesias speak of them, and Pliny says there were men in Ceylon who had long hairy tails, and were of remarkable swiftness of foot. Marco Polo tells us: "Now you must know that in this kingdom of Lambri[1] there are men with tails; these tails are of a palm in length, and have no hair on them. These people live in the mountains, and are a kind of wild men. Their tails are about the thickness of a dog's." Many modern travellers have heard of hairy and tailed people in the Malay Archipelago, and Mr. St. John, writing of Borneo, says that he met with a trader who had seen and felt the tails of a race which inhabited the north-east coast of the island. These tails were about four inches long, and so stiff that they had to use perforated seats. The Chinese also declare that in the mountains above Canton there is a race of tailed men. M. de Couret wrote about the Niam Niams, tailed men, who, he says, are living in Abyssinia or Nubia, having tails at least two inches long. We all know the old Lord Monboddo's theory that mankind had originally tails—nay, he went further, and said that some were born with them now—a fact which will be partially borne out by any military medical inspecting officer, who in the course of his practice has met with men whose "os coccygis" has been prolonged, so as to form a pseudo tail, which would unfit the man for the cavalry, although he would still be efficient as an infantry soldier.

Here is a very fine picture from a fresco at Pompeii representing tailed men, or, maybe, æsthetic young Fauns, treading out the vintage.

But tailed men are as nothing, compared to the wonderful beings that peopled the earth in bygone times.

---
[1] Supposed to be Sumatra.

It seems a pity that there are none of them now living, and that, consequent upon never having seen them, we are apt to imagine that they never existed, but were simply the creatures of the writer's brain. They were articles of belief until comparatively recent times, and

were familiar in Queen Elizabeth's time, as we learn from Othello's defence of himself (Act i. sc. 3) :—

> "And of the Cannibals that each other eat,
> The Anthropophagi, and men whose heads
> Do grow beneath their shoulders."

They were thoroughly believed in, a century or two

previously, in connection with Geography, and, in the "Mappa Mundi" (one of the earliest preserved English maps), now in Hereford Cathedral, which dates from the very early part of the fourteenth century, nearly the whole of the fanciful men hereafter mentioned are pourtrayed.

Sluper, who wrote in 1572, gives us the accompanying picture of a Cyclope, with the following remarks:—

> " De Polipheme & de Ciclopiens
> Tout mention Poetes anciens:
> On dit encor que ce lignage dure
> Auec vn oeil selon ceste figure."

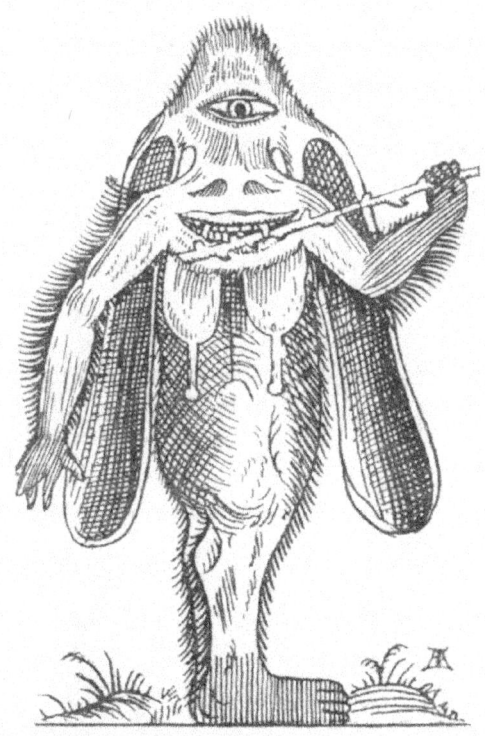

Pliny places the Cyclopes "in the very centre of the

earth, in Italy and Sicily;" and very likely there they might have existed, if we can bring ourselves to believe the very plausible explanation that they were miners, whose lanthorn, or candle, stuck in cap, was their one eye. At all events we may consider Sluper's picture as somewhat of a fancy portrait.

Among the Scythians, inhabiting the country beyond the Palus Mæotis, was a tribe which Herodotus (although he has been christened "The father of lies") did not believe in, nor indeed in any one-eyed men, but Pliny, living some 500 years after him, tells afresh the old story respecting these wonderful human beings. "In the vicinity also of those who dwell in the northern regions, and not far from the spot from which the north wind arises, and the place which is called its cave, and is known by the name of Geskleithron,[1] the Arimaspi are said to exist, a nation remarkable for having but one eye, and that placed in the middle of the forehead. This race is said to carry on a perpetual warfare with the Griffins,[2] a kind of monster, with wings, as they are commonly represented, for the gold which they dig out of the mines, and which these wild beasts retain, and keep watch over with a singular degree of cupidity, while the Arimaspi are equally desirous to get possession of it."

Milton mentions this tribe in "Paradise Lost," Book 2.

> "As when a Gryphon through the wilderness,
> With winged course, o'er hill, or mossy dale,
> Pursues the Arimaspian, who, by stealth,
> Had from his wakeful custody purloin'd
> The guarded gold."

---

[1] γης κλειθρον, meaning the limit or boundary of the earth.
[2] The Gryphon must not be confounded with the Griffin, as will be seen later on.

But there seems every probability that the story of the Gryphon was invented by the goldfinders, in order to deter people from coming near them, and interfering with their livelihood. There were, however, smaller Arimaspians, which probably the Gryphons did not heed, for Pliny tells us about the little thieves of mice. " In gold mines, too, their stomachs are opened for this purpose, and some of the metal is always to be found there, which they have pilfered, so great a delight do they take in stealing!" Livy, also, twice mentions mice gnawing gold.

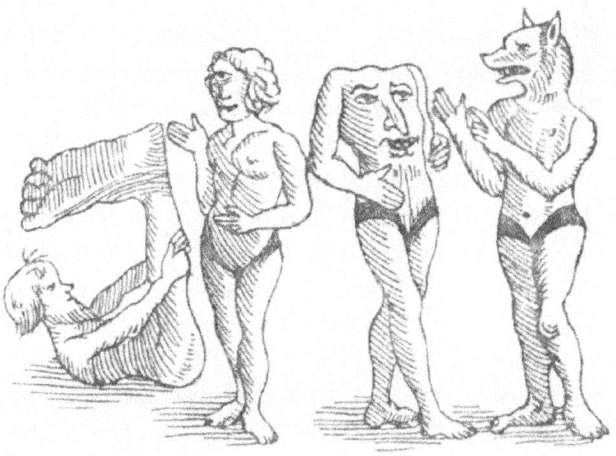

There were Anthropophagi—cannibals—as there are now, but, of course, they then lacked the luxury of cold missionary—and there were, besides, many wonderful beings. "Beyond the other Scythian Anthropophagi, there is a country called Abarimon, situate in a certain great valley of Mount Imaus (*the Himalayas*), the inhabitants of which are a savage race, whose feet are turned backwards, relatively to their legs; they possess wonderful velocity, and wander about indiscriminately with the wild beasts. We learn from Beeton, whose duty it was to take the measurements of the routes of Alexander the Great, that this people cannot breathe in any climate except their own, for which reason it is impossible to take them before any of the neighbouring kings; nor could any of them be brought before Alexander himself.

The Anthropophagi, whom we have previously mentioned as dwelling ten days' journey beyond the Borysthenes (*the Dneiper*), according to the account of Isogonus of Nicæa, were in the habit of drinking out of human skulls, and placing the scalps, with the hair attached, upon their breasts, like so many napkins. The same author relates that there is, in Albania, a certain race of men, whose eyes are of a sea-green colour, and who have white hair from their earliest childhood (*Albinos*), and that these people see better in the night than in the day. He states also that the Sauromatæ, who dwell ten days' journey beyond the Borysthenes, only take food every other day.

Crates of Pergamus relates, that there formerly existed in the vicinity of Parium, in the Hellespont (*Camanar, a town of Asia Minor*), a race of men whom he calls Ophiogenes, and that by their touch they were able to cure those who had been stung by serpents, extracting the poison by the mere imposition of the hand. Varro tells us, that there are still a few individuals in that district, whose saliva effectually cures the stings of serpents. The same, too, was the case with the tribe of the Psylli, in Africa, according to the account of Agatharcides; these people received their name from Psyllus, one of their kings, whose tomb is in existence, in the district of the Greater Syrtes

(*Gulf of Sidra*). In the bodies of these people, there was, by nature, a certain kind of poison, which was fatal to serpents, and the odour of which overpowered them with torpor; with them it was a custom to expose children, immediately after their birth, to the fiercest serpents, and in this manner to make proof of the fidelity of their wives; the serpents not being repelled by such children as were the offspring of adultery. This nation, however, was almost entirely extirpated by the slaughter made of them, by the Nasamones, who now occupy their territory. This race, however, still survives in a few persons, who are descendants of those who either took to flight, or else were absent on the occasion of the battle. The Marsi, in Italy, are still in possession of the same power, for which, it is said, they are indebted to their origin from the son of Circe, from whom they acquired it as a natural quality. But the fact is, that all men possess, in their bodies, a poison which acts upon serpents, and the human saliva, it is said, makes them take to flight, as though they had been touched with boiling water. The same substance, it is said, destroys them the moment it enters their throat, and more particularly so, if it should be the saliva of a man who is fasting.

Above the Nasamones (*living near the Gulf of Sidra*), and the Machlyæ, who border upon them, are found, as we learn from Calliphanes, the nation of the Androgyni, a people who unite the two sexes in the same individual, and alternately perform the functions of each. Aristotle also states, that their right breast is that of a male, the left that of a female.

Isigonus and Nymphodorus inform us that there are, in Africa, certain families of enchanters, who, by means of their charms, in form of commendations, can cause cattle to perish, trees to wither, and infants to die. Isigonus adds, that there are, among the Triballi, and the Illyrii, some persons of this description, who, also, have the power of fascination with the eyes, and can even kill those on whom they fix their gaze for any length of time, more especially if their look denotes anger: the age of puberty is said to be particularly obnoxious to the malign influence of such persons.

A still more remarkable circumstance is, the fact that these persons have two pupils in each eye. Apollonides says, that there are certain females of this description in Scythia, who are known as Bythiæ, and Phylarcus states that a tribe of the Thibii in Pontus, and many other persons as well, have a double pupil in one eye, and in the other the figure of a horse. He also remarks, that the bodies of these persons will not sink in water, even though weighed down by their garments. Damon gives an account of a race of people, not very much unlike them, the Pharnaces of Æthiopia, whose perspiration is productive of consumption to the body of every person that it touches. Cicero also, one of our own writers, makes the remark, that the glance of all women who have a double pupil is noxious.

To this extent, then, has nature, when she produced in man, in common with the wild beasts, a taste for human flesh, thought fit to produce poisons as well in every part of his body, and in the eyes of some persons, taking care that there shall be no evil influence in existence, which was not to be found in the human body. Not far from Rome, in the territory of the Falisci, a few families are found, who are known by the name of Hirpi. These people perform a yearly sacrifice to Apollo, on Mount Soracte, on which occasion they walk over a burning pile of wood, without being scorched even. On this account, by virtue of a decree of the Senate, they are always exempted from military service, and from all other public duties.

Some individuals, again, are born with certain parts of the body endowed with properties of a marvellous nature. Such was the case with King Pyrrhus, the great toe of whose right foot cured diseases of the spleen, merely by touching the patient. We are informed that this toe could not be reduced to ashes together with the other portions of his body; upon which it was placed in a temple.

India and the region of Æthiopia, more especially, abounds in wonders. In India the largest of animals are produced; their dogs, for instance, are much bigger than those of any other country. The trees, too, are said to be of such vast height that it is impossible to send an arrow over them. This is the result of the singular fertility of the soil, the equable temperature of the atmosphere, and the abundance of water; which, if we are to believe what is said, are such, that a single fig tree (*the banyan tree*) is capable of affording shelter to a whole troop of horse. The reeds here (*bamboos*) are of such enormous length, that each portion of them, between the joints, forms a tube, of which a boat is made that is capable of holding three men. It is a well-known fact, that many of the people here are more than five cubits in height.[1] These people never expectorate, are subject to no pains, either in the head, the teeth, and the eyes, and, rarely, in any other parts of the body;

---

[1] The Roman cubit was eighteen inches, so that these men were nearly eight feet high.

so well is the heat of the sun calculated to strengthen the constitution. . . . According to the account of Megasthenes, dwelling upon a mountain called Nulo, there is a race of men who have their feet turned backwards, with eight toes on each foot.

On many of the mountains again, there is a tribe of men who have the heads of dogs, and clothe themselves with the skins of wild beasts. Instead of speaking, they bark; and, furnished with claws, they live by hunting, and catching birds. According to the story, as given by Ctesias, the number of these people is more than a hundred and twenty thousand; and the same author tells us that there is a certain race in India, of which the females are pregnant once only in the course of their lives, and that the hair of the children becomes white the instant they are born. He speaks also of another race of men who are known as Monocoli,[1] who have only one leg, but are able to leap with surprising agility. The same people are also called Sciapodæ,[2] because they are in the habit of lying on their backs, during the time of extreme heat, and protect themselves from the sun by the shade of their feet. These people, he says, dwell not very far from the Troglodytæ (*dwellers in caves*); to the west of whom again there is a tribe who are without necks, and have eyes in their shoulders.[3]

Among the mountainous districts of the eastern parts of India, in what is called the country of the Cathareludi, we find the Satyr, an animal of extraordinary swiftness. These go sometimes on four feet, and sometimes walk erect; they have also the features of a human being. On account of their swiftness, these creatures are never

---

[1] From ἀπὸ τοῦ μονοῦ κώλου, "from having but one leg."
[2] From Σκιαποῦς, "making a shadow with his foot."
[3] See illustration, p. 9.

to be caught, except that they are aged, or sickly. Tauron gives the name of Choromandæ to a nation which dwells in the woods, and have no proper voice. These people screech in a frightful manner; their bodies are covered with hair, their eyes are of a sea-green colour, and their teeth like those of a dog. Eudoxus tells us, that in the southern parts of India, the men have feet a cubit in length, while the women are so remarkably small that they are called Struthpodes.[1]

Megasthenes places among the Nomades of India, a people who are called Scyritæ. These have merely holes in their faces instead of nostrils, and flexible feet, like the body of the serpent. At the very extremity of India, on the eastern side, near the source of the river Ganges, there is the nation of the Astomi, a people who have no mouths; their bodies are rough and hairy, and they cover themselves with a down[2] plucked from the leaves of trees. These people subsist only by breathing, and by the odours which they inhale through the nostrils. They support themselves neither upon meat nor drink; when they go upon a long journey they only carry with them various odoriferous roots and flowers, and wild apples, that they may not be without something to smell at. But an odour, which is a little more powerful than usual, easily destroys them. . . .

Isogonus informs us that the Cyrni, a people of India, live to their four-hundredth year; and he is of opinion that the same is the case also with the Æthiopian Macrobii,[3] the Seræ, and the inhabitants of Mount Athos. In the case of these last, it is supposed to be

---

[1] Sparrow footed, from στρούθος, a sparrow.
[2] Probably cotton.
[3] Or long livers, from μακρὸς, "long," and βίος, "life."

owing to the flesh of vipers, which they use as food; in consequence of which they are free also from all noxious animals, both in their hair and their garments.

According to Onesicritus, in those parts of India where there is no shadow, the men attain the height of five cubits and two palms,[1] and their life is prolonged to one hundred and thirty years; they die without any symptoms of old age, and just as if they were in the middle period of life. Pergannes calls the Indians, whose age exceeds one hundred years, by the name of Gymnetæ;[2] but not a few authors style them Macrobii. Ctesias mentions a tribe of them, known by the name of Pandore, whose locality is in the valleys, and who live to their two-hundredth year; their hair is white in youth, and becomes black in old age. On the other hand, there are some people joining up to the country of the Macrobii, who never live beyond their fortieth year, and their females have children once only during their lives. This circumstance is also mentioned by Agatharchides, who states, in addition, that they live on locusts, and are very swift of foot. Clitarchus and Megasthenes give these people the name of Mandi, and enumerate as many as three hundred villages which belong to them. Their women are capable of bearing children in the seventh year of their age, and become old at forty.

Artemidorus states that in the island of Taprobane

---

[1] A palm was three inches, so that these men would be eight feet high.
[2] From Γυμνητὴς, one who takes much bodily exercise.

CURIOUS CREATURES.

(*Ceylon*) life is prolonged to an extreme length, while at the same time, the body is exempt from weakness. Among the Calingæ, a nation also of India, the women conceive at five years of age, and do not live beyond their eighth year. In other places again, there are men born with long hairy tails, and of remarkable swiftness of foot; while there are others that have ears so large as to cover the whole body.

Crates of Pergamus states, that the Troglodytæ, who dwell beyond Æthiopia, are able to outrun the horse; and that a tribe of the Æthiopians, who are known as the Syrbotæ, exceed eight cubits in height (*twelve feet*). There is a tribe of Æthiopian Nomades dwelling on the banks of the river Astragus, towards the north, and about twenty days' journey from the ocean. These people are called Menismini; they live on the milk of the animal which we call cynocephalus (*baboon*), and rear large flocks of these creatures, taking care to kill the males, except such as they may preserve for the purposes of breeding. In the deserts of Africa, men are frequently seen to all appearance, and then vanish in an instant."[1]

It may be said that these descriptions of men are only the belief about the time of the Christian era, when Pliny lived—but it was the faith of centuries, and we find, 1200 years after Pliny died, Sir John Mandeville confirming his statements, and, as before stated, these wondrous creatures were given in illustrations, both in the Mappa Mundi, and in early printed books. Mandeville writes: " Many divers countreys & kingdoms are in Inde, and it is called Inde, of a river that runneth through it, which is called Inde also, and there are

[1] Mirage.

many precious stones in that river Inde. And in that ryver men finde Eles of xxx foote long, & men y^t dwell nere that river are of evill colour, yelowe & grene. . . .

"Then there is another yle that men call Dodyn, & it is a great yle. In this yle are maner diverse of men y^t have evyll maners, for the father eateth the son, & the son the father, the husband his wyfe, and the wyfe hir husbande. And if it so be that the father be sicke, or the mother, or any frend, the sonne goeth soone to the priest of the law & prayeth him that he will aske of the ydoll if his father shall dye of that sicknesse, or not. And then the priest and the son kneele down before the ydole devoutly, & asketh him, and he answereth to them, and if he say that he shall lyve, then they kepe him wel, and if he say that he shall dye, then commeth the priest with the son, or with the wyfe, or what frende that it be unto him y^t is sicke, and they lay their hands over his mouth to stop his breath, & so they sley him, & then they smite all the body into peces, & praieth all his frendes for to come and eate of him that is dead, and they make a great feste thereof, and have many minstrels there, and eate him with great melody. And so when they have eaten al ye flesh, then they take the bones, and bury them all singing with great worship, and all those that are of his frendes that were not at the eating of him, have great shame and vylany, so that they shall never more be taken as frends.

"And the king of this yle is a great lord and mightie, & he hath under him liii greate Yles, and eche of them hath a king; and in one of these yles are men that have but one eye, and that is in the middest of theyr front, and they eat flesh & fishe all rawe. And in another yle dwell men that have no heads, & theyr eyen are in theyr shoulders & theyr mouth is on theyr breste. In another yle are men that have no head ne eyen, and their mouth is in theyr shoulders. And in another yle are men that have flatte faces, without nose, and without eyen, but they have two small round holes in stede of eyen, and they have a flatte mouth without lippes. And in that yle are men that have their faces all flat without eyen, without mouth & without nose, but they have their eyen, and their mouth, behinde on their shoulders.

"And in another yle are foule men that have the lippes about the mouth so greate, that when they sleepe in the sonne they cover theyr face with the lippe. And in another yle are little men, as dwarfes, and have no mouth, but a lyttle rounde hole & through that hole they eate their meate with a pipe, & they have no tongue, & they speake not, but they blow & whistle, and so make signes one to another. And in another yle are wild men with hanging eares unto their shoulders. And in another yle are wild men, with hanging eares & have feete lyke an hors & they run faste, & they take wild beastes, and eate them. And in another yle are men that go on theyr handes & feete lyke beasts & are all rough, and will leape upon a tree like cattes or apes. And in another yle are men that go ever uppon theyr knees marvaylosly, and have on every foote viii Toes. . . .

"There is another yle that men call Pitan, men of this lande till no lande, for they eate nought, and they are smal, but not so smal as Pigmes. These men live with smell of wild aples, & when they go far out of the countrey, they beare apples with them, for anon, as they lose the savour of apples they dye—they are not reasonable, but as wyld beastes. And there is another yle where the people are all fethers,[1] but the face and the palmes of theyr handes, these men go as well about the sea, as on the lande, and they eate flesh and fish all raw. . . . In Ethiope are such men that have but one foote, and they go so fast y^t it is a great marvaill, & that is a large fote, that the shadow thereof covereth y^e body from son or rayne, when they lye upon their backes; and when their children be first borne they loke like russet, and when they waxe olde then they be all black."

There were also elephant-headed men.

In the olden times were men who did not build themselves houses—but sheltered themselves in caves, fissures of rocks, &c., and many are the remains we find of their flint implements, and the bones, which they used to split in order to extract the marrow of the animals they had slain with their rude flint arrows and spears. These, in classical times, were called Troglodytes (from the Greek τρωγλοδύται, *dwellers in caves*). It was a generic term, although particularly applied to uncivilised races on the banks of the Danube—those who dwelt on the

---

[1] Other editions read *rough hair*.

western coasts of the Red Sea—and Ethiopia. These latter could not have led a particularly happy life, for Herodotus tells us that the "Garamantes hunt the Ethiopian Troglodytes in four horse chariots; for the Ethiopian Troglodytes are the swiftest of foot of all men of whom we have heard any account given. The Troglodytes feed upon serpents and lizards, and such kind of reptiles; they speak a language like no other, but screech like bats."

Pliny, as we have seen, speaks of an adder eating people, whose food enables them to achieve extraordinary longevity, and Mandeville tells us that "From this yle, men go to an yle that is called Tracota, where all men are as beastes, & not reasonable, they dwell in caves, for they have not wyt to make them houses—they eate adders, and they speake not, but they make such a noyse as adders doe one to another, and they make no force of ryches, but of a stone that hath forty colours, and it is called Traconyt after that yle, they know not the vertue thereof, but they covete it for the great fayreness."

This stone was probably some kind of agate. It could not possibly have been a topaz, as some have thought, as the context from Pliny will show. "Topazos is a stone that is still held in very high estimation for its green tints; indeed, when first it was discovered, it was preferred to every other kind of precious stone. It so happened that some Troglodytic pirates, suffering from tempest and hunger, having landed upon an island off the coast of Arabia, known as Cytis, when digging there for roots and grass, discovered this precious stone; such, at least, is the opinion expressed by Archelaüs. Juba says that there is an island in the

Red Sea called *Topazos*, at a distance of three hundred stadia from the mainland; that it is surrounded by fogs, and is often sought by navigators in consequence; and that, to this, it received its present name, the word *Topazin*[1] meaning 'to seek' in the language of the Troglodytæ. . . . At a later period a statue, four cubits in height, was made of this stone. . . . Topazos is the largest of all the precious stones."

This shows that the Troglodytæ of Ethiopia had some commercial energy, and they did a good trade in myrrh and other condiments. Pliny says that the Troglodytæ traded among other things in cinnamon. They "after buying it of their neighbours, carry it over vast tracts of sea, upon rafts, which are neither steered by rudder nor drawn or impelled by oars or sails. Nor yet are they aided by any of the resources of art, man alone, and his daring boldness, standing in the place of all these; in addition to which, they choose the winter season, about the time of the equinox, for their voyage, for then a south-easterly wind is blowing; these winds guide them in a straight course from gulf to gulf, and after they have doubled the promontory of Arabia, the north-east wind carries them to a port of the Gebanitæ, known by the name of Ocilia. Hence it is that they steer for this port in preference, and they say that it is almost five years before the merchants are able to effect their return, while many perish on the voyage. In return for their wares, they bring back articles of glass and copper, cloths, buckles, bracelets, and necklaces; hence it is that this traffic depends more particularly upon the capricious tastes and inclinations of the female sex."

[1] In Greek, Τοπάζω, means to guess, divine, or conjecture.

This shows that some, at least, of the Troglodytes had a commercial spirit, and were in a comparative state of civilisation; in fact the latter is thoroughly proved, when, a little later on, Pliny speaks of Myrobalanum, "Among these various kinds, that which is sent from the country of the Troglodytæ is the worst of all," thus showing that they had reached the civilised pitch of adulteration! There are also several notices of peculiarities connected with this people, which deserve a passing glance. They had turtles with horns (or more probably fore-feet) which resembled the branches of a lyre; with these they swam. These were in all likelihood the tortoise-shell turtles, for they called them *Chelyon*. The Troglodytæ worshipped them. Their cattle were not like other oxen, for their horns pointed downwards to the ground, so that they were obliged to feed with their heads on one side. These oxen should have been crossed with those of Phrygia, whose horns were as mobile as their ears. And they were the happy possessors of a lake, called the *Unhealthy Lake*, which thrice a day became salt and bitter, and then again fresh, and this went on both day and night. We can hardly wonder that this *Lacus Insanus* was full of white serpents thirty feet long.

Amazons.

The race of Amazons or fighting women, is not yet extinct, as the chronicles of every police court can tell, and as an organised body of warlike soldiers—the King of Dahomey still keeps them up, or did until very recently. According to Herodotus, the Greeks, after having routed the Amazons, sailed away in three ships, taking with them as many Amazons, as they had been able to capture alive —but, when fairly out at sea, the ladies arose, stood up for women's rights, and cut all the Greeks in pieces. But they had not reckoned on one little thing, and that was, that none among them had the slightest idea of navigation; they couldn't even steer or row—so they had to drift about, until they came to Cremni (supposed to be near *Taganrog*), which was Scythian territory. They signalised their landing by horse-stealing, and the Scythians, not appreciating the joke, gave them battle, thinking they were men; but an examination of the dead proved them to be of the other sex. On learning this, the Scythians were far too gentlemanly to continue the strife, and, little by little, they established the most friendly relations with the Amazons. These ladies, however, objected to go to the Scythians' homes, for, as they pertinently put it, " We never could live with the women of your country, because we have not the same customs with them. We shoot with the bow, throw the javelin, and ride on horseback, and have never learnt the employments of women. But your women do none of the things we have mentioned, but are engaged in women's work, remaining in their wagons, and do not go out to hunt, or anywhere else; we could not therefore consort with them. If, then, you desire to have us for your wives, and to prove yourselves honest men, go to your parents, claim your share of their property, then return, and let us live by ourselves."

This the young Scythians did, but, when they returned, the Amazons said they were afraid to stop where they were, for they had deprived parents of their sons, and besides, had committed depredations in the country, so that they thought it but prudent to leave, and suggested that they should cross the Tanais, or *Don*, and found a colony on the other side. This their husbands acceded to, and when they were settled, their wives returned to their old way of living—hunting, going to war with their husbands, and wearing the same clothes—in fact they enjoyed an actual existence, of which many women nowadays, fondly, but vainly dream. There was a little drawback however—the qualification for a young lady's presentation at court, consisted of killing a man, and, until that was effected, she could not marry.

Sir John Mandeville of course knew all about them, although he does not pretend to have seen them, and this is what he tells us. "After the land of Caldee, is the land of Amazony, that is a land where there is no man but all women, as men say, for they wil suffer no man to lyve among them, nor to have lordeshippe over them. For sometyme was a kinge in that lande, and men were dwelling there as did in other countreys, and had wives, & it befell that the kynge had great warre with them of Sychy, he was called Colopius, and he was slaine in bataill and all the good bloude of his lande. And this Queene, when she herd that, & other ladies of that land, that the king and the lordes were slaine, they gathered them togither and killed all the men that were lefte in their lande among them, and sithen that time dwelled no man among them.

"And when they will have any man, they sende for them in a countrey that is nere theyr lande, and the men come, and are ther viii dayes, or as the woman lyketh, & then they go againe, and if they have men children they send them to theyr fathers, when they can eate & go, and if they have maide chyldren they kepe them, and if they bee of gentill bloud they brene[1] the left pappe[2] away, for bearing of a shielde, and, if they be of little bloud, they brene the ryght pappe away for shoting. For those women of that countrey are good warriours, and are often in soudy[3] with other lordes, and the queene of that lande governeth well that lande; this lande is all environed with water."

Pygmies.

The antitheses of men—Dwarfs, and Giants—must not be overlooked, as they are abnormal, and yet have existed in all ages. Dwarfs are mentioned in the Bible, *Leviticus* xxi. 20, where following the injunction of "Let him not approach to offer the bread of his God"—are mentioned the "crookbackt or dwarf." Dwarfs in all ages have been made the sport of Royalty, and the wealthy; but it is not of them I write, but of a race called the Pygmies, very small men who were descended from Pygmæus. They are noted in the earliest classics, for even Homer mentions them in his Iliad (B. 3, l. 3-6), which Pope translates:—

> "So, when inclement winter vex the plain
> With piercing frosts, or thick descending rain,
> To warmer seas, the Cranes embody'd fly,
> With noise, and order, through the mid-way sky;
> To pigmy nations, wounds and death they bring,
> And all the war descends upon the wing."

Homer also wrote a poem, "Pygmæogeranomachia,"

[1] Burn. [2] Breast. [3] At war.

about the Pygmies and Cranes. The accompanying illustration is from a fresco at Pompeii.

Aristotle says that they lived in holes under the earth, and came out in the harvest time with hatchets, to cut down the corn, as if to fell a forest, and went on goats and lambs of proportionable stature to themselves to make war against certain birds, called Cranes by

some, which came there yearly from Scythia to plunder them. Pliny mentions them several times, but especially in B. 7, c. 2. "Beyond these people, and at the very extremity of the mountains, the Trispithami,[1] and the

---

[1] From τρεις, *three*, σπιθαμαὶ, *spans*.

Pygmies are said to exist; two races, which are but three spans in height, that is to say, twenty-seven inches only. They enjoy a salubrious atmosphere, and a perpetual spring, being sheltered by the mountains from the northern blasts; it is these people that Homer has mentioned as being waged war upon by Cranes. It is said that they are in the habit of going down every spring to the sea-shore, in a large body, seated on the backs of rams and goats, and armed with arrows, and there destroy the eggs and the young of those birds; that this expedition occupies them for the space of three months, and that otherwise it would be impossible for them to withstand the increasing multitudes of the Cranes. Their cabins, it is said, are built of mud, mixed with feathers and egg shells."

Mandeville thus describes them. "When men passe from that citie of Chibens, they passe over a great river of freshe water, and it is nere iiii mile brode, & then men enter into the lande of the great Caan. This river goeth through the land of Pigmeens, and there men are of little stature, for they are but three span long, and they are right fayre, both men and women, though they bee little, and they live but viii[1] yeare, and he that liveth viii yeare is holden right olde, and these small men are the best workemen in sylke, and of cotton, in all maner of thing that are in the worlde; and these smal men travail not, nor tyl land, but they have amonge them great men, as we are, to travaill for them, & they have great scorne of those great men, as we would have of giaunts, or, of them, if they were among us."

Ser Marco Polo warns his readers against *pseudo* Pygmies. Says he: "I may tell you moreover that

[1] Other editions say, six or seven years.

when people bring over pygmies which they allege to come from India, 'tis all a lie and a cheat. For those little men, as they call them, are manufactured on this Island (*Sumatra*), and I will tell you how. You see there is on the Island a kind of monkey which is very small, and has a face just like a man's. They take these, and pluck out all the hair, except the hair of the beard, and on the breast, and then dry them, and stuff them, and daub them with saffron, and other things, until they look like men. But you see it is all a cheat; for nowhere in India, nor anywhere else in the world, were there ever men seen so small as these pretended pygmies."

But there are much more modern mention of these small folk. Olaus Magnus not only reproduces the classical story, but tells of the Pygmies of Greenland— the modern Esquimaux. These are also mentioned in Purchas his Pilgrimage, as living in Iceland, "pigmies represent the most perfect shape of man; that they are hairy to the uttermost joynts of the fingers, and that the males have beards downe to the knees; but, although they have the shape of men, yet they have little sense or understanding, nor distinct speech, but make shew of a kinde of hissing, after the manner of geese."

But to bring the history of pygmies down to modern times—I quote from "Giants and Dwarfs," by E. J. Wood, 1868, and I am thus particular in giving my authority, as the news comes from America, whence, sometimes, fact is mixed with fiction (pp. 246, 247, 248). "It is alleged by contemporary newspapers, that in 1828 several burying-grounds, from half an acre to an acre and a half in extent, were discovered in the county of White, state of Tennessee, near the town of Sparta, wherein very small people had been deposited in tombs or coffins of stone. The greatest length of the skeletons was nineteen inches. The bones were strong and well set, and the whole frames were well formed. Some of the people appeared to have lived to a great age, their teeth being worn smooth and short, while others were full and long. The graves were about two feet deep; the coffins were of stone, and made by laying a flat stone at the bottom, one at each side, or each end, and one over the corpse. The dead were all buried with their heads toward the east, and in regular order, laid on their backs, and with their hands on their breasts. In the bend of the left arm was found a cruse, or vessel, that would hold nearly a pint, made of ground stone, or shell, of a grey colour, in which were found two or three shells. One of these skeletons had about its neck ninety-four pearl beads. Near one of these burying-places was the appearance of the site of an ancient town.

Webber, in his 'Romance of Natural History,' refers to the diminutive sarcophagi found in Kentucky and Tennessee; and he describes these receptacles to be about three feet in length, by eighteen inches deep, and constructed, bottom, sides, and top, of flat, unhewn stones. These he conjectures to be the places of sepulture of a pigmy race, that became extinct at a period beyond reach even of the tradition of the so-called Indian aborigines.

Newspapers for 1866 tell us that General Milroy, who had been spending much time in Smith County, Tennessee, attending to some mining business, discovered near Watertown in that county some remarkable graves, which were disclosed by the washing of a small creek in its passage through a low bottom. The graves were from eighteen inches to two feet in length, most of them being of the smaller size, and were formed by an excavation of about fifteen inches below the surface, in which were placed four undressed slabs of rock—one in the bottom of the pit, one on each side, and one on the top. Human skeletons, some with nearly an entire skull, and many with well-defined bones, were found in them. The teeth were very diminutive, but evidently those of adults. Earthen crocks were also found with the skeletons. General Milroy could not gain any satisfactory information respecting these pigmy graves. The oldest inhabitants of the vicinity knew nothing of their origin or history, except that there was a large number of similar graves near Statesville in the same county, and also a little burial-ground at the mouth of Stone River, near the city of Nashville. General Milroy deposited the bones found by him in the State Library at Nashville."

That a race of dwarfs live in Central Africa, is now well known. Ronzo de Leo, who travelled in Africa, for many years with Dr. Livingstone, at one time almost stood alone in his assertion of this fact. But he was supported in his statement by G. Eugene Wolff, who had been in Central Africa with Stanley, and he maintained that, on the southern branches of the Congo, he had seen whole villages of Lilliputians, of whom the men were not over four and a half feet high, whilst the women were a great deal smaller. He described them as being both brave and cunning, expert with bow and arrow, with which they readily bring down the African bison, antelope, and even elephants. As trappers of small animals they are unsurpassed. In a close pinch they use the lance with astonishing dexterity, and an ordinary sling, in their hands, is wielded with wonderful skill.

These dwarfs collect the sap of the palm, with which they make soap. The men are smooth-faced, and of a rich mahogany colour, while the hair is short, and as black as night. Tens of thousands of them live on the south branch of the Congo.

Mr. Stanley in his expedition for the relief of Emin Pacha,[1] encountered some tribes of these pigmies, but he does not agree with the account which Mr. Wolff gives of them, who describes them as an affable, kind-hearted people, of simple ways, and devoid of vicious tendencies to a greater degree than most semi-barbaric races. The women are industrious and amiable.

Stanley, on the contrary, found them very annoying, and had a lively recollection of their poisoned arrows—but, at the present writing, he not having returned, and we, having no record but his letters, had better suspend our judgment as to the habits and tempers of these small people.

Wolff says they stand in awe of their bigger neighbours, but are so brave and cunning that, with all the odds of physique against them, the pigmies are masters of the situation.

Giants.

This last sentence seems almost a compendium of *The History of Tom Thumb*, for his wit enabled him to overcome the lubber-headed giants, in every conflict he was engaged in with them—they were no match for

---

[1] See his letters dated September 1888, which arrived in England early in April 1889.

him. Take the Romances of Chivalry. Pacolet, and all the dwarfs, were endowed with acute wits, and there was very little they could not compass—but the giants! their ultimate fate was always to be slain by some knight, and their imprisoned knights and damsels set free. A dwarf was a cleanly liver, but a giant was turbulent, quarrelsome, lustful, and occasionally cannibal. Fe Fi Fo Fum was the type of colossal man, and, as it is quite a pleasure to whitewash their characters in these respects, I hasten to do so before further discoursing on the subject of these great men.

It is Olaus Magnus who thus tells us

"Of the sobriety of Giants and Champions."

"That most famous Writer of the *Danish* affairs, *Saxo*, alleged before, and who shall be often alleged hereafter, saith, that amongst other mighty strong men in the *North*, who were as great as Giants, there was one *Starchaterus Thavestus*, whose admirable and heroick Vertues are so worthily extolled by him, that there were scarce any like him in those dayes in all *Europe*, or in the whole World, or hardly are now, or ever shall be. And amongst other Vertues he ascribes to that high-spirited man, he mentions his sobriety, which is principally necessary for valiant men : and I thought fit to annex that peculiarly to this relation, that we may, as in a glass, see more cleerly the luxury of this lustful age. For, as the same *Saxo* testifies, that valiant *Starchaterus* loved frugality, and loved not immoderate dainties. Alwayes neglecting pleasure, he respected Vertue, imitating the antient manner of Continency, and he desired a homely provision of his Diet ; he hated costly suppers ; wherefore hating profusion in Diet, and feeding on smoaked and rank meat, he drove away Hunger, with the greater appetite, as his meat was but of one kind, lest he should remit and abate the force of his true Vertue, by the contagion of outward Delights, as by some adulterate sweetness, or should abrogate the Rule of antient Frugality, by unusual Superstitions for Gluttony. Moreover, he could not endure to spend rost and boyled meat all at one Meal; holding that to be a monstrous Food, that Cookery had tampered with divers things together : Wherefore, that he might turn away the Luxury of the *Danes*, that they borrowed from the *Germans*, that made them so effeminate, amongst the rest he made Verses in his Country Language." Omitting many of them, he sang thus :

"*Starchaterus* his Verses on *Frugality*.

" Strong men do love raw meat ; nor do they need,
Or love, on dainty Cates and Feasts to feed,
War is the thing they most delight to breed,

CURIOUS CREATURES.   35

You may sooner bite off their beards that are
Full hard, and stiff with bristled, rugged, hair,
Than their wide mouths leave Milk their daily fare:
We fly from dainty Kitchins, and do fill
Our Bellies with rank Meats, and Countray swill,
Of old, men fed on boyl'd Meats, 'gainst their will.
A dish of Grass, that had no smack, did hold
Hog's and sheep's flesh together, hot or cold,
Nor to pollute their meats with mingling were they bold;
He that eats Cream we bid him for to be
Strong, and to have a mind that's bold and free.

.   .   .   .

Eleven Lords of elder time we were,
That waited on King Hachon, and at fare
Helgo Begachus sat first in order there.
First dish he eat was a dry'd Gammon, and
A Crust as hard as Flint he took in hand,
This made his hungry, yawning stomach stand:
No man at Table fed on stinking meat,
But what was good and common, each man eat,
Content with simple fare, though nere so great;
The greatest were not Gluttons, nor yet fine,
The King himself full sparingly would dine.
No Drinks were used that did of Honey bost,
Beer was their common Liquor, *Ceres* owest,
They fed on Meats were little boyl'd, no rost.
Each Table was with Meats but meanly drest,
Few Dishes on't, Antiquity thought best;
And in plain Fare each held himself most blest.
There were no Flagons, nor broad Bowls in use,
Nor painted Dishes grown to great abuse,
Each, at the Tap, did fill his wooden cruze.
No man, admirer of the former days,
Did use Tankards or Oxeys;[1] for their ways
Were sparing, almost empty Dishes this bewrays.
No Silver Basons, or guilt Cups were thought
Fit by the Host, and to the table brought,
To garnish, or by Ghests were vainly sought."

---

[1] Ox horns, horn cups.

By precept, and example, he induced many to Temperance and Sobriety—but, in spite of his moderation in food and drink, he was a most outrageous pirate, and Berserker.

At last, however, old, and weary of life, he sought death, and meeting Hatherus, son of a noble whom he had killed, begged him as a favour to cut his head off— and the young man, obligingly consenting, his head was severed from his body, and literally bit the ground. There are records of many more Northern giants, but none of so edifying a life as Starchaterus.

Giants are plentiful in the Bible, the Emins, Anakims, and the Zamzummims: there was Og, King of Bashan, whose iron bedstead was 9 cubits long by 4 broad— *i.e.*, 13 ft. 6 in. by 6 ft. That redoubtable champion of the Philistines, Goliath of Gath, was six cubits and a span high—*i.e.*, 9 ft. 9 in. In 2 Samuel xxi. 15–22, we find mention made of many giants.

"15 Moreover the Philistines had yet war again with Israel; and David went down, and his servants with him, and fought against the Philistines; and David waxed faint.

"16 And Ishbi-benob, which was of the sons of the giants, the weight of whose spear weighed three hundred shekels of brass in weight, he being girded with a new sword, thought to have slain David.

"17 But Abishai the son of Zeruiah succoured him, and smote the Philistine, and killed him. . . .

"18 And it came to pass after this, that there was again a battle with the Philistines at Gob: then Sibbechai the Hushathite slew Saph, which was of the sons of the giant.

"19 And there was again a battle in Gob with the

Philistines, where Elhanan the son of Jaare-oregim, a Bethlehemite, slew the brother of Goliath the Gittite, the staff of whose spear was like a weaver's beam.

"20 And there was yet a battle in Gath, where was a man of great stature, and on every foot six toes, four and twenty in number; and he also was born to the giant.

"21 And when he defied Israel, Jonathan the son of Shimeah, the brother of David, slew him.

"22 These four were born to the giant in Gath, and fell by the hand of David, and by the hand of his servants."

But these were mere pigmies if we can believe M. Henrion, who in 1718 calculated out the heights of divers notable persons—thus he found Adam was 121 ft. 9 in. high, Eve 118 ft. 9 in., Noah 27 ft., Abraham 20 ft., and Moses 13 ft.

Putting aside the mythical classical giants, Pliny says: "The tallest man that has been seen in our times, was one Gabbaras by name, who was brought from Arabia by the Emperor Claudius; his height was nine feet and as many inches. In the reign of Augustus, there were two persons, Posio and Secundilla, by name, who were half a foot taller than him; their bodies have been preserved as objects of curiosity in the Museum of the Sallustian family."

But it is reserved to Sir John Mandeville to have found the tallest giants of, comparatively speaking, modern times. "And beyond that valey is a great yle, where people as great as giaunts of xxviii fote long, and they have no clothinge but beasts skyns that hang on them, and they eate no bread, but flesh raw, and drink milke, and they have no houses, & they ate gladlyer fleshe of men, than other, & men saye to us that beyonde that yle is an yle where are greater giaunts as xlv or l fote long, & some said l cubits long (75 *feet*) but I saw them not, and among those giaunts are great shepe, and they beare great wolle, these shepe have I sene many times."

Early Men.

On the antiquity of man it is impossible to speculate, because we have no data to go upon. We know that his earliest existence, of which we have any cognisance, must have been at a period when the climate and fauna of the Western continent was totally different to their present state. Then roamed over the land, the elephant, rhinoceros, hippopotamus, the Bos-primigenius, the reindeer, the cave bear, the brown and the Arctic bears, the cave hyæna, and many other animals now quite extinct. We know that man then existed, because we find his handiwork in the shape of manufactured flint implements, mixed with the bones of these animals— and, occasionally, with them human remains have been found, but, as yet, no perfect skull has been found. There were two types of man, the Dolicho Cephalous, or long-headed, and the Brachy Cephalous, or round-headed—and, of these, the long-headed were of far greater antiquity.

All we can do is to classify man's habitation of this earth, as well as we can, under certain well-defined, and known conditions. Thus, that called the Stone Age, must be divided into two parts, that of the roughly chipped flint implements—which is designated the *Palæolithic* period—and that of the polished and carefully finished stone arms and implements, which necessarily show a later time, and a higher state of civilisation—which is called the *Neolithic* period. The next age is that of bronze, when man had learned to smelt metals, and make moulds, showing a great advance— and, finally, the Iron Age, in which man had subdued the sterner metal to his will—and this age immediately precedes History.

The cave men were of undoubted antiquity—and were hunters of the wild beasts that then overran Western Europe, and who split the bones of those animals which they slew in order to obtain the marrow. Although strictly belonging to the Palæolithic period, they manufactured out of that stubborn material, flint, spear-heads, knives, scrapers—and, when the bow had been invented, arrow-heads. Nor were they deficient in the rudiments of art, as some tracings and carvings on pieces of the horns of slaughtered animals, clearly show. Mr. Christie in digging in the Dordogne caves found, at La Madelaine, engraved and carved pictures of reindeer, an ibex, a mammoth, &c., all of them recognisable, and the mammoth, a very good likeness. This was incised on a piece of mammoth tusk.

The lake men, judging by the remains found near their dwellings, occupied their houses during the Stone and Bronze periods. Herodotus mentions these curious dwellings. "But those around Mount Pangæus and near the Doberes, the Agrianæ, Odomanti, and those who inhabit Lake Prasias[1] itself, were not at all subdued by Megabazus. Yet he attempted to conquer those who live upon the lake, in dwellings contrived after this manner: planks, fitted on lofty piles, are placed in the

[1] A lake between Macedonia and Thrace.

middle of the lake, with a narrow entrance from the mainland by a single bridge. These piles that support the planks, all the citizens anciently placed there at the common charge; but, afterwards, they established a law to the following effect; whenever a man marries, for each wife he sinks three piles, bringing wood from a mountain called Orbelus; but every man has several wives. They live in the following manner; every man has a hut on the planks, in which he dwells, with a trap door closely fitted in the planks, and leading down to the lake. They tie the young children with a cord round the foot, fearing lest they should fall into the lake beneath. To their horses and beasts of burden they give fish for fodder; of which there is such an abundance, that, when a man has opened his trap-door, he lets down an empty basket by a cord into the lake, and, after waiting a short time, draws it up full of fish."[1]

Here, then, we have a valuable record of the lake dwellings, and similar ones have been found in the lake of Zurich. In 1854, owing to the dryness and cold of the preceding winter, the water fell a foot below any previous record: and, in a small bay between Ober Meilen and Dollikon, the inhabitants took advantage to reclaim the soil thus left, and add it to their gardens, by building a wall as far out as they could—and they raised the level of the land thus gained, by dredging the mud out of the lake. In the course of dredging they found deer horns, tiles and various implements, and, the attention of an antiquary having been directed to this find, he concluded that it was the site of an ancient lake village. The lakes of Geneva, Constance, and

[1] The fishermen of lake Prasias still have lake dwellings as in the time of Herodotus.

Neufchatel, have also yielded much that throws light on the habits and intelligence of these lake men. They wove, they made pottery, they grew and parched corn—nay they ground it, and made biscuits, they ate apples, raspberries, blackberries, strawberries, hazel and beech nuts, and peas. They evidently fed on cereals, fruit, fish, and the flesh of wild animals, for bones of the following animals have been found. Brown bear, badger, marten, pine marten, polecat, wolf, fox, wild cat, beaver, elk, urus, bison, stag, roe-deer, wild boar, marsh boar—whilst their domestic animals were the boar, horse, ox, goat, sheep, and dog. These, it must be remembered, range over a wide period, including the stone and bronze ages. They wore ornaments, too, for pins, and bracelets have been found. Lake dwellings have been found in Scotland, England, Italy, Germany and France—so that this practice seems to have obtained very widely. In Ireland they made artificial islands in the lakes, called Crannoges, on which they erected their dwellings. Pile dwellings now exist, and are inhabited in many parts of the world.

We have other traces of prehistoric man in the shell mounds, kjökkenmöddings, or kitchen middens, which still exist in Denmark, and have been found in Scotland on the shores of the Moray Firth and Loch Spynie; in Cornwall, and Devon, at St. Valéry at the mouth of the Somme, in Australia, Tierra del Fuego, the Malay Peninsula, the Andaman Islands, and North and South America, showing a very wide range. The Danish kjökkenmöddings, when first thoroughly noticed, (of course, in this century), were taken to be raised beaches—but when they were examined, it was found that the shells were of four species of molluscs or shell-fish,[1] that did not live together, and that they were either full-grown, or nearly so. A stricter examination was made, and the result was the finding of some flint implements, and bones marked by knives, conclusively showing that man had had a hand in this collection of shells—and the conclusion was come to that these were the sites of villages of a prehistoric man, a hypothesis which was fully borne out by the discovery, in some of them, of hearths bearing traces of having borne fire. Thus, then, these refuse heaps were clearly the work of a very ancient race, so poor, and backward, as to be obliged to live on shell-fish—and these mounds were made by the shells which they threw away.

We can find a very great analogy between them and the Tierra del Fuegans, when Darwin visited them, while with the surveying ships *Adventure* and *Beagle*, a voyage which took from 1832 to 1836; and, when we read the following extracts from Darwin's account of the expedition, we can fancy we have before us a vivid picture of the makers of the kitchen middens. "The inhabitants, living chiefly upon shell-fish, are obliged constantly to change their place of residence; but they return at intervals to the same spots, as is evident from the pile of old shells, which must often amount to some tons in weight. These heaps can be distinguished at a long distance by the bright green colour of certain plants which invariably grow on them. . . . The Fuegian wigwam resembles, in size and dimensions, a haycock. It merely consists of a few broken branches stuck in the ground, and very imperfectly thatched on one side, with a few tufts of grass and rushes. The whole cannot be

[1] The most abundant were the oyster, mussel, cockle, and periwinkle.

so much as the work of an hour, and it is only used for a few days. . . . At a subsequent period, the *Beagle* anchored for a couple of days under Wollaston Island, which is a short way to the northward. While going on shore, we pulled alongside a canoe with six Fuegians. These were the most abject and miserable creatures I anywhere beheld. On the east coast, the natives, as we have seen, have guanaco cloaks, and, on the west, they possess sealskins. Amongst the central tribes the men generally possess an otter skin, or some small scrap about as large as a pocket handkerchief, which is barely sufficient to cover their backs as low down as their loins. It is laced across the breast by strings, and, according as the wind blows, it is shifted from side to side. But these Fuegians in the canoe were quite naked, and even one full-grown woman was absolutely so. It was raining heavily, and the fresh water, together with the spray, trickled down her body. . . . These poor wretches were stunted in their growth, their hideous faces bedaubed with white paint, their skins filthy and greasy, their hair entangled, their voices discordant, their gestures violent and without dignity. Viewing such men, one can hardly make oneself believe they are fellow-creatures and inhabitants of the same world. . . . At night, five or six human beings, naked, and scarcely protected from the wind and rain of this tempestuous climate, sleep on the wet ground, coiled up like animals. Whenever it is low water, they must rise to pick shellfish from the rocks; and the women, winter and summer, either dive and collect sea eggs, or sit patiently in their canoes, and, with a baited hair line, jerk out small fish. If a seal is killed, or the floating carcase of a putrid whale discovered, it is a feast: such miserable food is assisted by a few tasteless berries, and fungi. Nor are they exempt from famine, and, as a consequence, cannibalism accompanied by parricide."

This I believe to be as faithful a picture as can be drawn of the makers of the shell mounds.

But in Denmark, although shells formed by far the major part of these middens, yet they ate other fish, the herring, dorse, dab, and eel. Birds also were not despised by them, bones of swallows, the sparrow, stork, capercailzie, ducks, geese, wild swans, and even of the great auk (now extinct) have been found. Then of beasts they ate the stag, roe-deer, wild boar, urus, dog, fox, wolf, marten, otter, lynx, wild cat, hedgehog, bear, and mouse; beside which they lived on the seal, porpoise, and water rat.

Owing to the almost total absence of polished implements—and yet the fact being that portions of one or two have been found—the makers of these kjökkenmöddings, are classed as belonging to the later Palæolithic period.

Of the Bronze and Iron Ages there is no necessity to write, men were emerging from their primæval barbarity—and all the gentle arts, though undeveloped, were nascent. Men who could smelt metals, and mould, and forge them, cannot be considered as utter barbarians, such as were the long-headed men, with their chipped flint implements and weapons.

Wild Men.

Sometimes a specimen of humanity has got astray in infancy, and has been dragged up somehow in the woods, like Caspar Hauser, and Peter the Wild Boy, and fiction supplies other instances, such as Romulus and Remus, Orson, &c. Some of them were credited with being hairy as are the accompanying wild man and woman, as they are portrayed in John Sluper's book, where they are thus described:—

"L'Homme Sauvage.

"Combien que Dieu le createur seul sage,
A fait user les hommes de raison :
Icy voyez un vray homme sauvage,
Son corps vela est en toute saison."

"LA FEMME SAUVAGE.

"Femme sauvage a l'œil humain, non sainte,
Ainsi qu'elle est sur le naturel lieu,
Au naturel vous est icy depeinte,
Comme voyez qu'il appert a votre vue."

When Cæsar came to Britain for the second time, he found the Britons, although to a great extent civilised, having cavalry and charioteers (so many of the latter, that Cassivelaunus left about 4000 to watch the Romans), and knowing the art of fortification, yet in themselves, only just emerging from utter barbarism— the colouring and shaving of themselves showed that they had vanity, and were making, after their fashion, the most of their personal charms. Cæsar (Book v. 14) writes: "Of all these *tribes*, by far the most civilised are those who inhabit Kent, which district is altogether maritime; nor do they differ much from the Gallic customs. Most of those in the interior do not sow corn, but live on flesh and milk, and are clad in skins. All the Britons, in truth, dye themselves with woad, which produces a bluish colour, and on this account they are of a more frightful aspect in battle. They have flowing hair, and every part of the body shaved, except the head and the upper lip. Ten, and *even* twelve of them have wives in common between them, and chiefly brothers with brothers, and fathers with sons; but, if there is any offspring, they are considered to be the children of those by whom each virgin was first espoused."

Hairy Men.

If, as we may conjecture from the above, the ancient Briton was "a rugged man, o'ergrown with hair," his full-dress toilette must have occupied some time. But extreme hairiness in human beings is by no means singular, and very many cases are recorded in medical books. Many of us may remember the Spanish dancer, Julia Pastrana, whose whole body was hairy, and who had a fine beard. She had a child on whom the hair began to grow, like its mother; and, but a few years back, there was a hairy family exhibited in London — their faces being covered with hair, as is the case of the *Puella pilosa*, or Hairy Girl — given by Aldrovandus in his *Monstrorum Historia*.

She was aged twelve years, and came from the Canary Isles, together with her father (aged 40), her brother (20),

48 CURIOUS CREATURES.

and her sister (8), all as hairy one as the other. They were brought over by Marius Casalius, and first shown at Bologna, so that this is no doubt a faithful likeness,

as Aldrovandus lived and died in that city. He gives other examples, but not so well authenticated as this.

There were two wonderful hairy people at Ava, in Burmah, who are described by two most trustworthy eye-witnesses, John Crawford, in his "Journal of an Embassy from the Governor-General of India to the Court of Ava"—and in 1855, by Captain Henry Youle, in his "Narrative of the Mission sent by the Governor-General of India to the Court of Ava." They were father and daughter, respectively named Shu-Maon, and Maphoon. The father may strictly be said to have had neither eyelashes, eyebrows, nor beard, because the whole of his face, including the interior and exterior of his ears, were covered with long silky silvery grey hair. His whole body, except his hands and feet, was covered with hair of the same texture and colour as that now described, but generally less abundant; it was most plentiful over the spine and shoulders, where it was five inches long; over the breast, about four inches, and was most scanty on the arms, legs, thighs, and abdomen.

Of the daughter, Captain Youle writes: "The whole of Maphoon's face was more or less covered with hair. On a part of the cheek, and between the nose and mouth, this was confined to a short down, but over all the rest of the face was a thick silky hair of a brown colour, paleing about the nose and chin, four or five inches long. At the alæ of the nose, under the eye, and on the cheek bone this was very fully developed; but it was in, and on, the ear, that it was most extraordinary. Except the upper tip, no part of the ear was visible. All the rest was filled and veiled with a large mass of silky hair, growing apparently out of every part of the external organ, and hanging a pendant lock to a length of eight or ten inches. The hair over her forehead was brushed so as to blend with the hair of the head, the latter being dressed (as usual with her countrywomen) *à la Chinoise;* it was not so thick as to conceal her forehead.

"The nose, densely covered with hair, as no animal's is, that I know of, and with long locks curving out, and pendant like the wisps of a fine Skye-terrier's coat, had a most strange appearance. The beard was pale in colour, and about four inches in length, seemingly very soft and silky."

Maphoon, when Captain Youle saw her, had two children, one, the eldest, perfectly normal, the other, who was very young, was evidently taking after its mother.

The Aïnos, an aboriginal tribe in the north of Japan, who are looked down upon by the Japanese as dogs, have always been reputed as being covered with hair. Mr. W. Martin Wood read a paper before the Ethnological Society of London [1] respecting them, and he said, " Esau himself could not have been a more hairy man than are these Aïnos. The hair forms an enormous bush, and it is thick and matted. Their beards are very thick and long, and the greater part of their face is covered with hair which is generally dark in colour; they have prominent foreheads, and mild, dark eyes, which somewhat relieve the savage aspect of their visage. Their hands and arms, and, indeed, the greater part of their bodies, are covered with an abnormal profusion of hair."

This, however, has been questioned, notably by Mr. Barnard Davis, whose paper may be read in the 3rd vol. of the "Memoirs of the Anthropological Society of London"—and he quotes from several travellers, to prove that the hairyness of the Aïnos had been exaggerated. However, Miss Bird in her "Unbeaten Tracks in Japan" may fairly be said to have put the subject at rest, for

[1] Transactions of the Ethnological Society, 1866, vol. iv., p. 34.

she visited, and travelled in the Aïno country. She, certainly, disproves the theory that, as a race, they were hairy, although she confesses that some were—as, for instance (p. 232), "They wore no clothing, but only one was hairy," and, writing from Biratori, Yezo (p. 255), she says, "The men are about the middle height, broad-chested, broad-shouldered, thick set, very strongly built, the arms and legs short, thick, and muscular, the hands and feet large. The bodies, and especially the limbs of many, are covered with short, bristly hair. I have seen two boys whose backs are covered with fur as fine, and soft, as that of a cat." Again (p. 283), "The profusion of black hair, and a curious intensity about their eyes, coupled with the hairy limbs and singularly vigorous *physique*, give them a formidably savage appearance; but the smile, full of 'sweetness and light,' in which both eyes and mouth bear part, and the low, musical voice, softer and sweeter than anything I have previously heard, make me, at times, forget that they are savages at all."

The Ouran Outan.

Transition from hirsute humanity to the apes, is easy, and natural—and we need only deal with the Simiinæ, which includes the Orang, the Chimpanzee, and the Gorilla. These are the largest apes, and nearest approach to man—but, although they may be tailless, yet there is that short great toe which prevents any acceptation of their humanity. The orang is exclusively an inhabitant of Borneo and Sumatra, and in those two islands it may be found in the swampy forests near the coast. It grows to a large size, for an ape, about four feet four inches high, but is neither so large, nor so strong, as the Gorilla. Compared with man, its arms seem to be as extravagantly long, as its legs are ridiculously short. When wild, it feeds entirely on vegetable diet, and makes a kind of house, or nest, in trees, interweaving the branches, so as to obtain shelter. They do not stand confinement well, being languid and miserable—but, in their native wildness, they can, if necessity arises, fight well in their own defence. A. R. Wallace, in his "Malay Archipelago; the Land of the Orang Utan and the Bird of Paradise," tells the following story of its combativeness.

"A few miles down the river there is a Dyak house, and the inhabitants saw a large orang feeding on the young shoots of a palm by the river side. On being alarmed, he retreated towards the jungle, which was close by, and a number of the men, armed with spears and choppers, ran out to intercept him. The man who was in front, tried to run his spear through the animal's body, but the orang seized it in his hands, and in an instant got hold of the man's arm, which he seized in his mouth, making his teeth meet in the flesh above the elbow, which he tore and lacerated in a dreadful manner. Had not the others been close behind, the man would have been seriously injured, if not killed, as he was quite powerless; but they soon destroyed the creature with their spears and choppers. The man remained ill for a long time, and never fully recovered the use of his arm."

It is called the Simia Satyrus; probably on its presumed lustfulness, certainly not on account of its resemblance to the satyr of antiquity.

Gesner gives us his idea of the orang, presenting us with the accompanying figure of the Cercopithecus, and quotes Cardanus as saying that the Cercopithecus or Wild-man, is singularly made, having the height and form of a man, with legs like man's—and is covered all over with hair. No animal can withstand it, with the exception of man, to whom, when in its own regions, it is not inferior. It loves boys and women.

Pliny speaks of the Satyr Ape thus: "Among the mountainous districts of the eastern parts of India, in what is called the country of the Catharcludi, we find the Satyr, an animal of extraordinary swiftness. They go sometimes on four feet, and sometimes walk erect; they have, also, the features of a human being. On account of their swiftness, these creatures are never to be caught, except when they are aged, or sickly," and, in another place, he says, "The Sphyngium and the Satyr stow away food in the pouches of their cheeks, after which they will take out piece by piece in their hands, and eat it."

Topsell has mixed up the Simia Satyrus with the classical satyr, having legs and horns like goats; but he evidently alludes to the former in this passage. "The

Satyres are in the Islands *Satiridæ*, which are three in number, right over against India on the farther side of the *Ganges;* of which *Euphemus Car* rehearseth this history: that when he sailed unto *Italy*, by the rage of winde and evill weather, they were driven to a coast unnavigable, where were many desart Islandes, inhabited of wild men, and the marriners refused to land upon some Islands, having heretofore had triall of the inhumaine and uncivill behaviour of the inhabitants, so that they brought us to the *Satyrian Islands*, where we saw the inhabitants red, and had tayles joyned to their backs, not much lesse than horsses. These, being perceived by the marriners to run to the shippes, and lay hold on the women that were in them, the shipmen, for feare, took one of the Barbarian women, and set her on the land among them, whom in most odious and filthy manner, they abused, whereby they found them to be very bruit beasts."

He gives us his idea of the Simia Satyrus, which must have been an accomplished animal, for not only could it, apparently, play upon the pipe, but it had a handy pouch for the reception of the fruit (in lieu of coppers) which it doubtless would receive as guerdon for its performance.

Satyrs.

He also mentions and delineates a curious Ape which closely resembles the classical Satyr: "Under the *Equinoctiall*, toward the East and South, there is a kind of Ape called *Ægopithecus*, an Ape like a Goate. For there are Apes like Beares, called *Arctopitheci*, and some like Lyons, called *Leontopitheci*, and some like Dogs, called *Cynocephali*, as is before expressed; and many other which have a mixt resemblance of other creatures in their members.

"Amongst the rest there is a beast called PAN; who in his head, face, horns, legs, and from the loynes downward resembleth a Goat, but in his belly, breast, and armes, an Ape: such a one was sent by the King of *Indians* to Constantine, which, being shut up in a cave or close place, by reason of the wildnesse thereof, lived there but a season, and when it was dead and bowelled, they pouldred it with spices, and carried it to be seene at Constantinople: the which beast having beene seene of the ancient Græcians, were so amazed at the strangenesse thereof, that they received it for a God, as they did a Satyre, and other strange beasts."

I have said that Topsell has mixed the Ape and the Satyr, inextricably—but as his version has the charm of description and anecdote, I give it with little curtailment.

"As the *Cynocephali*, or *Baboun* Apes have given occasion to some to imagine (though falsly) there were such men, so the *Satyre*, a most rare and seldom seene beast, hath occasioned other to thinke it was a Devil; and the Poets with their Apes, the Painters, Limners, and Carvers, to encrease that superstition, have therefore described him with hornes on his head, and feet like Goates, whereas Satires have neither of both. And it may be that Devils have at some time appeared to men in this likenes, as they have done in the likeness of the *Onocentaure* and wild Asse, and other shapes; it being also probable that Devils take not any dænomination or shape from Satyres, but rather the Apes themselves, from Devils whom they resemble, for there are many things common to the Satyre Apes, and devilish Satyres, as their human shape, their abode in solitary places, their rough hayre, and lust to women, wherewith all other Apes are naturally infected; but especially Satyres. . . .

"Peradventure the name of Satyre is more fitly derived from the Hebrew, *Sair. Esa.* 34, whereof the plural is *Seirim, Esa.* 13, which is interpreted monsters of the Desart, or rough hairy Fawnes; and when *Iisim* is put to *Seir*, it signifieth Goates.

"The *Chaldæans*, for *Seirim*, render *Schedin;* that is, evill devills; and the *Arabians*, *lesejathin*, that is *Satanas:* the *Persyans*, *Devan*, the *Illyrians*, *Devadai*, and *Dewas:* the *Germans*, *Teufel*. They which passed through the world, and exercised daunting and other sports for *Dionisius*, were called Satyres, and sometimes *Tytiri*, because of their wanton songes; sometimes *Sileni* (although the difference is, that the smaller and younger beasts are called *Satiri*, the elder, and greater, *Sileni;*) Also *Bacchæ* and *Nymphæ*, wherefore *Bacchus* is pictured riding in a chariot of vine branches, *Silenus* ridinge beside him on an Asse, and the *Bacchæ* or *Satyres* shaking togetheer their staulkie Javelines and Paulmers.[1] By reason of their leaping they are called *Scirti*, and the anticke or satyrical dauncing, *Sicinnis*, and they also sometimes *Sicinnistæ*; sometimes *Ægipanæ*; wherefore *Pliny* reporteth, that among the westerne *Ethiopians*, there are certain little hilles full of the *Satirique Ægipanæ*, and that, in the night-time they use great fires, piping and dansing, with a wonderful noise of Tymbrels and Cymbals; and so also in *Atlas* amongest the Moores, whereof there was no footing, remnant, or appearance, to be found in the daytime.

" . . . There are also *Satires* in the Eastern mountaines of *India*, in the country of the *Cartaduli*, and in the province of the *Comari* and *Corudæ*, but the *Cebi* spoken of before, bred in *Ethiopia*, are not *Satyres* (though faced like them :) nor the *Prasyan* Apes, which resemble *Satyres* in short beards. There are many kindes of these *Satyres* better distinguished by names than any properties naturall known unto us. Such are the *Ægipanæ*, before declared, *Nymphes* of the Poets, *Fawnes, Pan* and *Sileni*, which, in time of the Gentiles were worshipped for Gods; and it was one part of their religion to set up the picture of a Satyre at their dores and gates, for a remedy against the bewitching of envious persons.

" . . . Satyres have no humaine conditions in them, nor any other resemblance of men besides their outward shape; though *Solinus* speakes of them like as of men. They carry their meate under their chin as in a store house, and from thence being hungry, they take it forth to eat, making it ordinary with them every day, which is but annuall in the *Formicæ* lions; being of very unquiet

[1] Thyrsi.

motions above other Apes. They are hardly taken, except sicke, great with yong, old or asleepe; for *Sylla* had a *Satyre* brought him, which was taken asleepe neare *Apollonia*, in the holy place *Nymphæum*, of whom he (by divers interpreters) demanded many questions, but received no answer, save only a voice very much like the neighing of a horse, wherof he being afraid, sent him away alive.

"*Philostratus* telleth another history, how that *Apollonius* and his colleagues, supping in a village of *Ethiopia*, beyond the fall of *Nilus*, they heard a sudden outcry of women calling to one another; some saying, *Take him*, others, *Follow him;* likewise provoking their husbands to helpe them: the men presently tooke clubs, stones, or what came first to hand, complaining of an injury done unto their wives. Now some ten moneths before, there had appeared a fearfull shew of a Satyre, raging upon their women, and had slain two of them, with whom he was in love: the companions of *Apollonius* quaked at the hearing hereof, and *Nilus*, one of them, swore (by *Jove*) that they being naked and unarmed, could not be able to resist him in his outragious lust, but that he would accomplish his wantonnes as before: yet, said *Apollonius*, there is a remedy to quaile these wanton-leaping beasts, which men say *Midas* used (for *Midas* was of kindred to *Satyres*, as appeared by his eares). This *Midas* heard his mother say, that *Satyres* loved to be drunke with wine, and then sleep soundly, and after that, be so moderate, mild and gentle, that a man might thinke they had lost their first nature.

"Whereupon he put wine into a fountain neere the highway, whereof, when the *Satyre* had tasted, he waxed meeke suddenly, and was overcome. Now that we thinke not this a fable (saith *Apollonius*) let us go to the Governor of the Towne, and inquire of him whether there be any wine to be had that we may offer it to the *Satyre*, wherunto all consented, and they filled foure great *Egyptian* earthen vessels with wine, and put it in the fountain where their cattel were watred : this done, *Apollonius* called the *Satyre*, secretly thretning him, and the *Satire*, inraged with the savour of the wine came ; after he had drunke thereof, Now, said *Apollonius*, let us sacrifice to the *Satyre*, for he sleepeth, and so led the inhabitants to the dens of the *Nymphs*, distant a furlong from the towne, and shewed them the *Satyre* saying ; Neither beat, curße, or provoke him henceforth, and he shall never harme you.

"It is certaine, that the devills do many waies delude men in the likeness of *Satyres ;* for, when the drunken feasts of *Bacchus* were yearely celebrated in *Parnassus*, there were many sightes of *Satyres*, and voyces, and sounding of cymbals heard : yet it is likely that there are men also like Satyres, inhabiting in some desart places ; for *S. Ierom*, in the life of *Paul the Eremite*, reporteth that there appeared to *S. Anthony*, an *Hippocentaure* such as the Poets describe, and presently he saw, in a rocky valley adjoining, a little man having croked nostrils, hornes growing out of his forhed, and the neather part of his body had Goat's feet ; the holy man, not dismayed, taking the shield of faith, and the breastplate of righteousnesse, like a good souldior of Christ, pressed toward him, which brought him some fruites of palmes as pledges of his peace, upon which he fed in the journey ; which Saint *Anthony* perceiving, he asked him who he was, and received this answere ; I am a mortall creature, one of the inhabitants of this Desart, whom the Gentiles (deceived with error) doe worship, and call *Fauni*, *Satyres*, and *Incubi :* I am come in ambassage from our flocke, intreating that thou would'st pray for us unto the common GOD, who came to save the world; the which words were no sooner ended, but he ran away as fast as any foule could fly. And least this should seeme false, under *Constantine* at *Alexandria* there was such a man to be seene alive, and was a publick spectacle to all the World; the carcasse thereof, after his death, was kept from corruption by heat, through salt, and was carried to *Antiocha* that the Emperor himself might see it.

"*Satyres* are very sildom seene, and taken with great difficulty, as is before saide: for there were two of these founde in the woods of *Saxony* towards *Dacia*, in a desart, the female was killed by the darts of the hunters, and the biting of Dogs, but the male was taken alive, being in the upper parts like a man, and in the neather partes like a Goat, but all hairy throughout: he was brought to be tame, and learned to go upright, and also to speake some wordes, but with a voice like a Goat, and without all reason.

"The famous learned man *George Fabricius*, shewed me this shape of a monstrous beast that is fit to be joyned to the story of *Satyres*. There was, (saide he,) in the territory of the Bishop of *Salceburgh*, in a forrest called *Fannesbergh*, a certaine foure-footed beast, of a yellowish carnation colour, but so wilde that he would never be drawne to looke upon any man, hiding himselfe in the darkest places, and beeing watched diligently, would not be provoked to come forth so much as to eate his meate —so that in a very short time it was famished. The hinder legs were much unlike the former, and also much longer. It was taken about the year of the Lord, one thousand five hundred, thirty, whose image being here so lively described, may save us further labour in discoursing of his maine and different parts and proportion."

The Sphynx.

"The Sphynga or *Sphinx*, is of the kind of Apes, but his breast up to his necke, pilde and smooth without hayre:

the face is very round, yet sharp and piked, having the beasts of women, and their favor, or visage, much like them : In that part of the body which is bare with out haire, there is a certaine red thing rising in a round circle, like millet seed, which giveth great grace &  comeliness to their coulour, which in the middle part is humaine : Their voice is very like a man's, but not articulate, sounding as if one did speake hastily, with indignation or sorrow. Their haire browne, or swarthy coulour. They are bred in *India*, and *Ethiopia*. In the promontory of the farthest *Arabia* neere *Dira*, are *Sphinges*, and certaine *Lyons*, called *Formicæ*, so, likewise, they are to be found amongest the *Trogloditæ*.

"As the *Babouns* and *Cynocephali* are more wilde than other Apes, so the *Satyres* and *Sphynges* are more meeke and gentle, for they are not so wilde that they will not bee tamed, nor yet so tame, but they will revenge their own harmes ; as appeared by that which was slayne in a publike spectacle among the *Thebanes*. They carrye their meat in the store houses of their own chaps or cheeks, taking it forth when they are hungry, and so eat it.

"The name of this *Sphynx* is taken from 'binding,' as appeareth by the Greek notation, or else of delicacie and dainty nice loosnesse, (wherefore there were certain common strumpets called *Sphinctæ*, and the *Megarian Sphingas* was a very popular phrase for notorious harlots), hath given occasion to the poets to faigne a certaine

monster called *Sphynx*, which they say was thus derived.

Hydra brought foorth the *Chimæra*, *Chimæra* by *Orthus*, the *Sphynx*, and the *Nemæan* Lyon: now, this *Orthus* was one of *Geryon's* dogges. This *Sphynx* they make a treble formed monster, a Mayden's face, a Lyon's legs, and the wings of a fowle; or, as *Ansonius* and *Varinus* say, the face and head of a mayde, the body of a dogge, the winges of a byrd, the voice of a man, the clawes of a Lyon, and the tayle of a dragon : and that she kept continually in the *Sphincian* mountaine; propounding to all travailers that came that way an *Ænigma*, or Riddle, which was this : *What was the creature that first of all goeth on foure legges; afterwards on two, and, lastly, on three:* and all of them that could not dissolve that Riddle, she presently slew, by taking them, and throwing them downe headlong, from the top of a Rocke. At last *Œdipus* came that way, and declared the secret, that it *was a man, who in his infancy creepeth on all foure*, afterward, *in youth, goeth upon two legs*, and last of all, *in olde age taketh unto him a staffe which maketh him to goe, as it were, on three legs;* which the monster hearing, she presently threwe down herselfe from the former rocke, and so she ended. Whereupon Œdipus is taken for a subtill and wise opener of mysteries.

"But the truth is, that when *Cadmus* had married an *Amazonian* woman, called *Sphynx*, and, with her, came to *Thebes*, and there slew *Draco* their king, and possessed his kingdom, afterwards there was a sister unto *Draco* called *Harmona*, whom *Cadmus* married, *Sphynx* being yet alive. She, in revenge, (being assisted by many followers,) departed with great store of wealth into the mountaine *Sphincius*, taking with her a great Dogge, which *Cadmus* held in great account, and there made daily incursions or spoiles upon his people. Now, *ænigma*, in the *Theban* language, signifieth an inrode, or warlike incursion, wherfore the people complained in this sort. *This* Grecian Sphinx *robbeth us, in setting up with an* ænigma, *but no man knoweth after what manner she maketh this* ænigma.

"*Cadmus* hereupon made proclamation, that he would give a very bountifull reward unto him that would kill *Sphinx*, upon which occasion the Corinthian *Œdipus* came unto her, being mounted on a swift courser, and accompanied with some *Thebans* in the night season, slue her. Other say that *Œdipus* by counterfaiting friendshippe, slue her, making shew to be of her faction; and *Pausanius* saith, that the former Riddle, was not a Riddle, but an Oracle of *Apollo*, which *Cadmus* had received, whereby his posterity should be inheritors of the *Theban* kingdome; and whereas *Œdipus*, being the son of *Laius*, a former king of that countrey, was taught the Oracle in his sleepe, he recouvered the kingdome usurped by *Sphinx* his sister, and, afterwards, unknown, married his mother Jocasta.

"But the true morall of this poetical fiction is by that learned *Alciatus*, in one of his emblems, deciphered; that her monstrous treble formed shape signified her lustfull pleasure under a Virgin's face, her cruell pride, under the Lyon's clawes, her winde-driven leuitye, under the Eagles, or birdes feathers, and I will conclude with the wordes of *Suidas* concerning such monsters, that the *Tritons, Sphinges*, and *Centaures*, are the images of those things, which are not to be founde within the compasse of the whole world."

Apes.

Sluper, who could soar to the height of delineating a Cyclops, is equal to the occasion when he has to deal with Apes, and here he gives us an Ape which, unfortunately, does not seem to have survived to modern times—namely, one which wove for itself coarse cloth, probably of rushes; had a cloak of skin, and walked upright, with the aid of a walking-stick, and was so genteel, that, having no boots, he seems to have blacked his feet. And thus he sings of it:

> " Pres le Peru par effect le voit on,
>   Dieu a donné au Singe telle forme.
> Vestu dejonc, s'appuyant d'un baston,
>   Estāt debout, chose aux hōmes cōforme."

Before quitting the subject of Apes, I cannot refrain from noticing another of this genus mentioned by Topsell, and that is the Arctopithecus or Bear Ape:—"There is in America a very deformed beast, which the inhabitants call *Haut* or *Hauti*, and the Frenchmen *Guenon*, as big as a great Affrican Monkey. His belly hangeth very low, his head and face like unto a childes, and being taken, it will sigh like a young childe. His skin is of an ashe-colour, and hairie like a Beare: he hath but three clawes on a foote, as longe as foure fingers, and like the thornes of Privet, whereby he climbeth up into the highest trees, and for the most part liveth of the leaves of a certain tree, beeing of an exceeding heighth, which the *Americans* call *Amahut*, and thereof this beast is called *Haut*. Their tayle is about three fingers long, having very little haire thereon; it hath beene often tried, that though it suffer any famine, it will not eate the fleshe of a living man, and one of them was given me by a French-man, which I kept alive sixe and twenty daies, and at the last it was killed by Dogges, and in that time when I had set it abroad in the open ayre, I observed that, *although it often rained, yet was that beast never wet.*[1] When it is tame, it is very loving to a man, and desirous to climbe uppe to his shoulders, which those naked *Amerycans* cannot endure, by reason of the sharpnesse of his Clawes."

Animal Lore.

We are indebted to Pliny for much strange animal lore—which, however, will scarcely bear the fierce light of modern investigation. Thus, he tells us of places in which certain animals are not to be found, and narrates some very curious zoological anecdotes thereon. "It is a remarkable fact, that nature has not only assigned different countries to different animals, but that even in the same country it has denied certain species to certain localities. In Italy, the dormouse is found in one part only, the Messian forest. In Lycia, the gazelle never passes beyond the mountains which border upon Syria;

[1] The italics are mine.—J. A.

nor does the wild ass in that vicinity pass over those which divide Cappadocia from Cilicia. On the banks of the Hellespont, the stags never pass into a strange territory, and, about Arginussa, they never go beyond Mount Elaphus; those upon the mountains, too, have cloven ears. In the island of Poroselene, the weasels will not so much as cross a certain road. In Bœotia, the moles, which were introduced at Lebadea, fly from the very soil of that country, while in the neighbourhood, at Orchomenus, the very same animals tear up all the fields. We have seen coverlets for beds made of the skin of these creatures, so that our sense of religion does not prevent us from employing these ominous animals for the purposes of luxury.

"When hares have been brought to Ithaca, they die as soon as ever they touch the shore, and the same is the case with rabbits, on the shores of the island of Ebusus; while they abound in the vicinity, Spain namely, and the Balearic isles. In Cyrene, the frogs were formerly dumb, and this species still exists, although croaking ones were carried over there from the Continent. At the present day, even, the frogs of the island of Seriphos are dumb; but when they are carried to other places, they croak; the same thing is also said to have taken place at Sicandrus, a lake of Thessaly. In Italy, the bite of a shrew-mouse is venomous; an animal which is not to be found in any region beyond the Apennines. In whatever country it exists, it always dies immediately if it goes across the rut made by a wheel. Upon Olympus, a mountain of Macedonia, there are no wolves, nor yet in the isle of Crete. In this island there are neither foxes nor bears, nor, indeed, any kind of baneful animal, with the exception of the phalangium, a species of spider. It is a thing still more remarkable, that in this island there are no stags, except in the district of Cydon; the same is the case with the wild boar, the woodcock, and the hedgehog."

He further tells us of animals which will injure strangers only, as also animals which injure the natives only.

"There are certain animals which are harmless to the natives of the country, but destroy strangers; such as the little serpents at Tirynthus, which are said to spring out of the earth. In Syria, also, and especially on the banks of the Euphrates, the serpents never attack the Syrians when they are asleep, and even if they happen to bite a native who treads upon them, their venom is not felt; but to persons of any other country they are extremely hostile, and fiercely attack them, causing a death attended with great torture. On this account the Syrians never kill them. On the contrary, on Latmos, a mountain of Caria, as Aristotle tells us, strangers are not injured by the scorpions, while the natives are killed by them."

He also throws some curious light, unknown to modern zoologists, on the antipathies of animals one to another. He says:—"There will be no difficulty in perceiving that animals are possessed of other instincts besides those previously mentioned. In fact, there are certain antipathies, and sympathies among them, which give rise to various affections, besides those which we have mentioned in relation to each species, in its appropriate place. The Swan and the Eagle are always at variance, and the Raven and the Chloreus seek each other's eggs by night. In a similar manner, also, the Raven and the Kite are perpetually at war with one another, the one carrying off the other's food. So, too, there are antipathies between the Crow and the Owl, the Eagle and the Trochilus; between the last two, if we are to believe the story, because the latter has received the title of 'the king of birds;' the same, again, with the Owlet and all the smaller birds.

"Again, in relation to the terrestrial animals, the Weasel is at enmity with the Crow, the Turtle-dove with the Pyrallis, the Ichneumon with the Wasp, and the Phalangium with other Spiders. Among aquatic animals, there is enmity between the Duck and the Seamew, the Falcon known as the 'Harpe,' and the Hawk called the 'Triorchis.' In a similar manner, too, the Shrewmouse and the Heron are ever on the watch for each other's young; and the Ægithus, so small a bird as it is, has an antipathy for the Ass; for the latter, when scratching itself, rubs its body against the brambles, and so crushes the bird's nest; a thing of which it stands in such dread, that, if it only hears the voice of the Ass when it brays, it will throw its eggs out of the nest, and the young ones, themselves, will, sometimes, fall to the ground in their fright; hence it is that it will fly at the Ass, and peck at its sores with its beak.

"The Fox, too, is at war with the Nisus, and Serpents with Weasels and Swine. Æsalon is the name given to a small bird that breaks the eggs of the Raven, and the young of which are anxiously sought by the Fox; while, in its turn, it will peck at the young of the Fox, and even the parent itself. As soon as the Ravens espy this, they come to its assistance, as though against a common enemy. The Acanthis, too, lives among the brambles; hence it is that it also has an antipathy to the Ass, because it devours the bramble blossoms. The

Ægithus and the Anthus, too, are at such mortal enmity with each other, that it is the common belief that their blood will not mingle; and it is for this reason that they have the bad repute of being employed in many magical incantations. The Thos and the Lion are at war with each other; and, indeed, the smallest objects and the greatest, just as much. Caterpillars will avoid a tree that is infested with Ants. The Spider, poised in its web, will throw itself on the head of a Serpent, as it lies stretched beneath the shade of the tree where it has built, and, with its bite, pierce its brain; such is the shock, that the creature will hiss from time to time, and then, seized with vertigo, coil round and round, while it finds itself unable to take to flight, or so much as to break the web of the spider, as it hangs suspended above; this scene only ends with its death."

The Manticora.

Of curious animals, other than Apes, depicted as having some approach to the human countenance, perhaps the most curious is the Manticora. It is not a *parvenu*; it is of ancient date, for Aristotle mentions it. Speaking of the dentition of animals, he says :—" None of these genera have a double row of teeth. But, if we may believe Ctesias, there are some which have this peculiarity, for he mentions an Indian animal called Martichora, which had three rows of teeth in each jaw; it is as large and rough as a lion, and has similar feet, but its ears and face are like those of a man; its eye is grey, and its body red; it has a tail like a land Scorpion, in which there is a sting; it darts forth the spines with which it is covered, instead of hair, and it utters a noise resembling the united sound of a pipe and a trumpet; it is not less swift of foot than a stag, and is wild, and devours men."

Pliny also quotes Ctesias, but he slightly diverges, for he says it has azure eyes, and is of the colour of blood; he also affirms it can imitate the human speech. *Par parenthèse* he mentions, in conjunction with the Manticora, another animal similarly gifted:—" By the union of the hyæna with the Æthiopian lioness, the Corocotta is produced, which has the same faculty of imitating the voices of men and cattle. Its gaze is always fixed and immoveable; it has no gums in either of its jaws, and the teeth are one continuous piece of bone; they are enclosed in a sort of box, as it were, that they may not be blunted by rubbing against each other."

Mais, revenons à nos moutons, or rather Mantichora. Topsell, in making mention of this beast, recapitulates all that Ctesias has said on the subject, and adds:— " And I take it to be the same Beast which *Avicen* calleth *Marion,* and *Maricomorion,* with her taile she woundeth her Hunters, whether they come before her or behinde her, and, presently, when the quils are cast forth, new ones grow up in their roome, wherewithal she overcometh all the hunters; and, although India be full of divers ravening beastes, yet none of them are stiled with a title of *Andropophagi,* that is to say, Men-eaters; except onely this *Mantichora.* When the Indians take a Whelp of this beast, they fall to and bruise the buttockes and taile thereof, so that it may never be fit to bring (*forth*) sharp quils, afterwards it is tamed without peril. This, also, is the same beast which is called *Leucrocuta,* about the bignesse of a wilde Asse, being in legs and hoofes like a Hart, having his mouth reaching on

both sides to his eares, and the head and face of a female like unto a Badgers. It is also called *Martiora*, which in the Parsian tongue, signifieth a devourer of men."

Du Bartas, in "His First Week, or the Birth of the World," mentions our friend as being created :—

> "Then th' *Vnicorn*, th' *Hyæna* tearing tombs,
> Swift *Mantichor*', and *Nubian Cephus* comes ;
> Of which last three, each hath, (as heer they stand)
> Man's voice, Man's visage, Man like foot and hand."

It is mentioned by other writers—but I have a theory of my own about it, and that is, that it is only an idealised laughing hyæna.

The Lamia.

The Lamiæ are mythological—and were monsters of Africa, with the face and breast of a woman, the rest of the body like that of a serpent; they allured strangers, that they might devour them ; and though not endowed with the faculty of speech, their hissings were pleasing. Some believed them to be evil spirits, who, in the form of beautiful women, enticed young children, and devoured them; according to some, the fable of the Lamiæ is derived from the amours of Jupiter with a beautiful woman, Lamia, whom Juno rendered deformed, and whose children she destroyed ; Lamia became insane, and so desperate, that she ate up all the children which came in her way.

Topsell, before entering upon the natural history of the Lamia, as an animal, tells the following story of it as a mythological being :—"It is reported of *Menippus* the Lycian, that he fell in love with a strange woman, who at that time seemed both beautifull, tender, and rich, but, in truth, there was no such thing, and all was but a fantastical ostentation; she was said to insinuate her selfe, into his familiaritie after this manner: as he went upon a day alone from *Corinth* to *Senchræa*, hee met with a certaine phantasme, or spectre like a beautifull woman, who tooke him by the hand, and told him she was a *Phœnician* woman, and of long time had loved him dearely, having sought many occasions to manifest the same, but could never finde opportunitie untill that day, wherefore she entreated him to take knowledge of her house, which was in the Suburbes of *Corinth*, therewithall pointing unto it with her finger, and so desired his presence. The young man seeing himselfe thus wooed by a beautiful woman, was easily overcome by her allurements, and did oftimes frequent her company.

"There was a certaine wise man, and a Philosopher, which espied the same, and spake unto *Menippus* in this manner, 'O formose, et a formorsis, expetitie mulieribus, ophin thalpies, cai se ophis,' that is to say, 'O fair *Menippus*, beloved of beautiful women, art thou a serpent, and dost nourish a serpent?' by which words he gave him his first admonition, or incling of a mischiefe; but not prevayling, *Menippus* proposed to marry with this spectre, her house to the outward shew, being richly furnished with all manner of houshold goods; then said the wise man againe unto *Menippus*, 'This gold, silver, and ornaments of house, are like to *Tantalus* Apples, who are said by *Homer* to make a faire shew, but to containe in them no substance at all; even so, whatsoever you conceave of this riches, there is no matter or substance in the things which you see, for they are onely inchaunted images, and shadowes, which that you may beleeve, this your neate bride is one of the *Empusæ*, called *Lamia*, or *Mormolicæ*, wonderfull desirous of commerce with men, and loving their flesh above measure; but those whom they doe entice, afterwards they devoure without love or pittie, feeding upon their flesh.' At which words the wise man caused the gold and silver plate, and household stuffe, cookes, and servants to vanish all away. Then did the spectre like unto one that wept, entreate the wise man that he would not torment her, nor yet cause her to confesse what manner of person she was; but he on the other side being inexorable, compelled her to declare the whole truth, which was, that she was a Phairy, and that she purposed to use the companie of *Menippus*, and feede him fat with all manner of pleasures, to the extent that, afterward, she might eate up and devour his body, for all their kinde love was only to feed upon beautiful yong men. . . .

"To leave therefore these fables, and come to the true description of the *Lamia*, we have in hand. In the foure and thirty chapter of Esay, we do find this called a beast *Lilith* in the Hæbrew, and translated by the auncients *Lamia*, which is threatened to possesse *Babell*. Likewise in the fourth chapter of the Lamentations, where it is said in our English translation, that the Dragons lay forth their brests, in Hæbrew they are called *Ehannum*, which, by the confession of the best interpreters, cannot signifie Dragons, but rather Sea calves, being a generall word for strange wilde beasts. How be it the matter being wel examined, it shall appeare that it must needes be this Lamia, because of her great breastes, which are not competible either to the Dragon, or Sea calves; so then, we wil take it for graunted, by the testimony of holy Scripture, that there is such a beast as this *Cristostinius*. *Dion* also writeth that there are such beasts in some parts of *Libia*, having a Woman's face, and very beautifull, also very large and comely shapes on their breasts, such as cannot be counterfeited by the art of any painter, having a very excellent colour in their fore parts, without wings, and no other voice but hissing like Dragons: they are the

swiftest of foote of all earthly beasts, so as none can escape them by running, for, by their celerity, they compasse their prey of beastes, and by their fraud they overthrow men. For when they see a man, they lay open their breastes, and by the beauty thereof, entice them to come neare to conference, and so, having them within their compasse, they devoure and kill them.

"Unto the same things subscribe *Cælius* and *Giraldus*, adding also, that there is a certaine crooked place in

Libia neare the Sea-shore, full of sand like to a sandy Sea, and all the neighbor places thereunto are deserts. If it fortune at any time, that through shipwrack, men come there on shore, these beasts watch uppon them, devouring them all, which either endevour to travell on the land, or else to returne backe againe to Sea, adding also, that when they see a man they stand stone still, and stir not til he come unto them, looking down upon their breasts or to the ground, whereupon some have thought, that seeing them, at their first sight have such a desire to come neare them, that they are drawne into their compasse, by a certaine naturall magicall witchcraft. . . . The hinderparts of the beast are like unto a Goate, his fore legs like a Beares, his upper parts to a woman, the body scaled all over like a Dragon, as some have affirmed by the observation of their bodies, when *Probus*, the Emperor, brought them forth unto publike spectacle; also it is reported of them, that they devoure their own young ones, and therefore they derive their name *Lamia*, of *Lamiando*; and thus much for this beast."

The Centaur.

This extraordinary combination of man and animal is very ancient—and the first I can find is Assyrian. Mr. W. St. Chad Boscawen, in one of his British Museum Lectures (afterwards published under the title of *From under the Dust of Ages*), speaking of the seasons and the zodiacal signs, in his lecture on *The Legend of Gizdhubar*, says:—" Gizdhubar has a dream that the stars of heaven are falling upon him, and, like Nebuchadnezzar, he can find no one to explain the hidden meaning to him. He is, however, told by his huntsman, Zaidu, of a very wise creature who dwells in the marshes, three days' journey from Erech. . . . The strange being, whom this companion of the hero is despatched to bring to the Court, is one of the most interesting in the Epic. He is called Hea-bani—'he whom Hea has made.' This mysterious creature is represented on the gems, as half a man, and half a bull. He has the body, face, and arms of a man, and the horns, legs, hoofs, and tail of a bull. Though in form rather resembling the satyrs, and in fondness for, and in association with the cattle, the rustic deity Pan, yet in his companionship with Gizdhubar, and his strange death, he approaches nearer the Centaur Chiron, who was the companion of Heracles.

"By his name he was the son of Hea, whom Berosus identifies as Cronos, as Chiron was the son of Cronos. Like Chiron, he was celebrated for his wisdom, and acted as the counsellor of the hero, interpreting his dreams, and enabling him to overcome the enemies who attacked him. Chiron met his death at the hand of Heracles, one of whose poisoned arrows struck him, and, though immortal, he would not live any longer, and gave his immortality to Prometheus. . . . Zeus made Chiron among the stars a Sagittarius. Here again we have a striking echo of the Chaldæan legend, in the Erech story. According to the arrangement of tablets, the death of Hea-bani takes place under the sign of Sagittarius, and is the result of some fatal accident during the combat between Gizdhubar and Khumbaba. Like the Centaurs, before his call to the Court of Gizdhubar, Hea-bani led a wild and savage life. It is said on the tablets 'that he consorted with the wild beasts. With the gazelles he took his food by night, and consorted with the cattle by day, and rejoiced his heart with the creeping things of the waters.'

"Hea-Bani was true and loyal to Gizdhubar, and when Istar (the Assyrian Venus), foiled in her love for Gizdhubar, flew to heaven to see her father Anu (the Chaldæan Zeus), and to seek redress for the slight put upon her, the latter created a winged bull, called 'The Bull of Heaven,' which was sent to earth. Hea-Bani, however, helps his lord, the bull is slain, and the two companions enter Erech in triumph. Hea-Bani met with his death when Gizdhubar fought Khumbaba, and 'Gizdhubar for Hea-Bani his friend wept bitterly and lay on the ground.'"

Thus, centuries before the Romans had emerged from barbarism, we have the prototype of the classical Centaur, the man-horse. The fabled Centaurs were a people of Thessaly—half-men, half-horses—and their existence is very cloudy. Still, they were often depicted, and the two examples of a male and female Centaur, from a fresco at Pompeii, are charmingly drawn. It will be seen that both are attended by Bacchantes bearing thyrses—a delicate allusion to their love of wine; for it was owing to this weakness that their famous battle with the Lapithæ took place. The Centaurs were invited to the marriage of Hippodamia with Pirithous, and, after the manner of cow-boys "up town," they got intoxicated, were very rude, and even offered violence to the women present. That, the good knights, Sir Hercules and Sir Theseus, could not stand, and with the Lapithæ, gave the Centaurs a thrashing, and made them retire to Arcadia. They had a second fight over the matter of wine, for the Centaur Pholus gave Hercules to drink of wine meant for him, but in the keeping of the Centaurs, and these ill-conditioned animals resented it, and attacked Hercules with fury. They were fearfully punished, and but few survived.

Pliny pooh-poohs the mythical origin of the Centaurs, and says they were Thessalians, who dwelt along Mount

Pelion, and were the first to fight on horseback. Aldrovandus writes that, according to Licosthenes, there were formerly found, in the regions of the Great Tamberlane, Centaurs of such a form as its upper part was that of a man, with two arms resembling those of a toad,

and he gives a drawing from that author, so that the reader might diligently meditate whether such an animal was possible in a natural state of things; but the artist seems to have forgotten the fore-legs.

"The Onocentaur is a monstrous beast;
Supposed halfe a man, and halfe an Asse,
That never shuts his eyes in quiet rest,
Till he his foes deare life hath round encompast.
Such were the Centaures in their tyrannie,
That liv'd by Humane flesh and villanie."
—CHESTER.

THE GORGON.

In the title-page of one edition of "The Historie of Foure-footed Beastes" (1607) Topsell gives this picture of the Gorgon; and he says, respecting this curious animal, the following:—"Among the manifold and divers sorts

84   CURIOUS CREATURES.

of Beasts which are bred in Affricke, it is thought that
the *Gorgon* is brought foorth in that countrey. It is a
feareful and terrible beast to behold : it hath high and
thicke eie-lids, eies not very great, but much like an
Oxes or Bugils, but all fiery bloudy, which neyther looke
directly forwarde, nor yet upwards, but continuallye downe
to the earth, and therefore are called in Greeke *Catoble-
ponta*. From the crowne of their head downe to their
nose, they have a long hanging mane, which makes them

to look fearefully. It eateth deadly and poysonfull hearbs,
and if at any time he see a Bull, or other creature whereof
he is afraid, he presently causeth his mane to stand upright, and, being so lifted up, opening his lips, and gaping
wide, sendeth forth of his throat a certaine sharpe and
horrible breath, which infecteth, and poysoneth the air
above his head, so that all living creatures which draw
the breath of that aire are greevously afflicted thereby,
loosing both voyce and sight, they fall into leathall and

deadly convulsions. It is bred in *Hesperia* and *Lybia*.

"The Poets have a fiction that the *Gorgones* were the Daughters of *Medusa* and *Phorcynis*, and are called *Steingo*, and by *Hesiodus*, *Stheno*, and *Eyryale* inhabiting the *Gorgadion* Ilands in the *Æthiopick Ocean*, over against the gardens of *Hesperia*. *Medusa* is said to have the haires of his head to be living Serpentes, against whom *Perseus* fought, and cut off his hed, for which cause he was placed in heaven on the North side of the *Zodiacke* above the Waggon, and on the left hand holding the *Gorgons* head. The truth is that there were certaine *Amazonian* women in *Affricke* divers from the *Scythians*, against whom *Perseus* made warre, and the captaine of those women was called *Medusa*, whom *Perseus* overthrew, and cut off her head, and from thence came the Poet's fiction describing Snakes growing out of it as is aforesaid. These *Gorgons* are bred in that countrey, and have such haire about their heads, as not onely exceedeth all other beastes, but also poysoneth, when he standeth upright. Pliny calleth this beast *Catablepon*,[1] because it continually looketh downwards, and saith all the parts of it are but smal excepting the head, which is very heavy, and exceedeth the proportion of his body, which is never lifted up, but all living creatures die that see his eies.

"By which there ariseth a question whether the poison which he sendeth foorth, proceede from his breath, or from his eyes. Whereupon it is more probable, that like the Cockatrice, he killeth by seeing, than by the breath of his mouth, which is not competible to any other beasts in the world. Besides, when the Souldiers

[1] From καταβλέπω, "to look downwards."

of *Marius* followed *Iugurtha*, they saw one of these *Gorgons*, and, supposing it was some sheepe, bending the head continually to the earth, and moving slowly, they set upon him with their swords, whereat the Beast, disdaining, suddenly discovered his eies, setting his haire upright, at the sight whereof the Souldiers fel downe dead.

"*Marius*, hearing thereof, sent other souldiers to kill the beaste, but they likewise died, as the former. At last the inhabitantes of the countrey, tolde the Captaine the poyson of this beast's nature, and that if he were not killed upon a Sodayne, with onely the sight of his eies he sent death into his hunters: then did the Captaine lay an ambush of souldiers for him, who slew him sodainely with their speares, and brought him to the Emperour, whereupon *Marius* sent his skinne to Rome, which was hung up in the Temple of *Hercules*, wherein the people were feasted after the triumphes; by which it is apparent that they kill with their eies, and not with their breath. . . .

"But to omit these fables, it is certaine that sharp poisoned sightes are called *Gorgon Blepen*, and therefore we will followe the Authoritie of *Pliny* and *Athenæus*. It is a beast set all over with scales like a Dragon, having no haire except on his head, great teeth like Swine, having wings to flie, and hands to handle, in stature betwixt a Bull and a Calfe.

"There be Ilandes called *Gorgonies*, wherein these monster-*Gorgons* were bredde, and unto the daies of *Pliny*, the people of that countrey retained some part of their prodigious nature. It is reported by *Xenophon*, that *Hanno*, King of *Carthage*, ranged with his armie in that region, and founde there, certaine women of incredible swiftenesse and perniscitie of foote. Whereof he tooke two onely of all that appeared in sight, which had such roughe and sharp bodies, as never before were seene. Wherefore, when they were dead, he hung up their skinnes in the Temple of *Juno*, for a monument of their straunge natures, which remained there untill the destruction of *Carthage*. By the consideration of this beast, there appeareth one manifest argument of the Creator's devine wisdome and providence, who hath turned the eies of this beaste downeward to the eartd, as it were thereby burying his poyson from the hurt of man ; and shaddowing them with rough, long and strong haire, that their poysoned beames should not reflect upwards, untill the beast were provoked by feare or danger, the heavines of his head being like a clogge to restraine the liberty of his poysonfull nature, but what other partes, vertues or vices, are contained in the compasse of this monster, God onely knoweth, who, peradventure, hath permitted it to live uppon the face of the earth, for no other cause but to be a punishment and scourge unto mankind ; and an evident example of his owne wrathfull power to everlasting destruction. And this much may serve for a description of this beast, untill by God's providence, more can be known thereof."

The Unicorn.

What a curious belief was that of the Unicorn ! Yet what mythical animal is more familiar to Englishmen ? In its present form it was not known to the ancients, not even to Pliny, whose idea of the Monoceros or Unicorn is peculiar. He describes this animal as having "the head of a stag, the feet of an elephant, the tail of the boar, while the rest of the body is like that of the horse : it makes a deep lowing noise, and has a single black horn, which projects from the middle of its forehead, two cubits in length. This animal, it is said, cannot be taken alive."

Until James VI. of Scotland ascended the English throne as James I., the Unicorn, as it is now heraldically portrayed (which was a supporter to the arms of James IV.) was almost unknown—vide *Tempest*, iii. 3. 20 :—

"*Alonzo*. Give us kind keepers, heavens : what were these ?
Sebastian. A living drollery. Now I will believe that there are unicorns."

Spenser, who died before the accession of James I., and therefore did not write about the supporters of the Royal Arms, alludes (in his *Faerie Queene*) to the antagonism between the Lion and the Unicorne.

" Likë as the lyon, whose imperial poure
A proud rebellious unicorn defyes,
T'avoide the rash assault, and wrathful stoure
Of his fiers foe, him to a tree applyes,
And when him rouning in full course he spyes,
He slips aside : the whiles that furious beast,
His precious horne, sought of his enimyes,
Strikes in the stroke, ne thence can be released,
But to the victor yields a bounteous feast."

Pliny makes no mention of the Unicorn as we have it heraldically represented, but speaks of the Indian Ass, which, he says, is only a one-horned animal. Other old naturalists, with the exception of Ælian, do not mention it as our Unicorn—and his description of it hardly coincides. He says that the Brahmins tell of the wonderful beasts in the inaccessible regions of the interior of India, among them being the Unicorn, " which they call *Cartazonon*, and say that it reaches the size of a horse of mature age, possesses a mane and reddish-yellow hair, and that it excels in swiftness through the excellence of its feet and of its whole body.

Like the elephant it has inarticulate feet, and it has a boar's tail; one black horn projects between the eyebrows, not awkwardly, but with a certain natural twist, and terminating in a sharp point."

Guillim, who wrote on heraldry in 1610, gives, in his Illustrations, indifferently the tail of this animal, as horse or ass; and, as might be expected from one of his craft, magnifies the Unicorn exceedingly:—"The Unicorn hath his Name of his one Horn on his Forehead. There is another Beast of a huge Strength and Greatness, which hath but one Horn, but that is growing on his Snout, whence he is called *Rinoceros*, and both are named *Monoceros*, or *One horned*. It hath been much questioned among Naturalists, which it is that is properly called the Unicorn: And some hath made Doubt whether there be any such Beast as this, or no. But the great esteem of his Horn (in many places to be seen) may take away that needless scruple. . . .

"Touching the invincible Nature of this Beast, *Job* saith, '*Wilt thou trust him because his Strength is great, and cast thy Labour unto him? Wilt thou believe him, that he will bring home thy seed, and gather it into thy Barn?*' And his Vertue is no less famous than his Strength, in that his Horn is supposed to be the most powerful Antidote against Poison: Insomuch as the general Conceit is, that the wild Beasts of the Wilderness use not to drink of the Pools, for fear of the venemous Serpents there breeding, before the Unicorn hath stirred it with his Horn. Howsoever it be, this Charge may very well be a Representation both of Strength or Courage, and also of vertuous Dispositions and Ability to do Good; for to have Strength of Body, without the Gifts and good Qualities of the Mind, is but the Property of an Ox, but where both concur, that may truly be called Manliness. And that these two should consort together, the Ancients did signify, when they made this one Word, *Virtus*, to imply both the Strength of Body, and Vertue of the Mind. . . .

"It seemeth, by a Question moved by *Farnesius*, That the Unicorn is never taken alive; and the Reason being demanded, it is answered 'That the greatness of his Mind is such, that he chuseth rather to die than to be taken alive: Wherein (saith he) the Unicorn and the valiant-minded Souldier are alike, which both contemn Death, and rather than they will be compelled to undergo any base Servitude or Bondage, they will lose their Lives.' . . .

"The Unicorn is an untameable Beast by Nature, as

may be gathered from the Words of *Job, chap.* 39, '*Will the Unicorn serve thee, or will he tarry by thy Crib? Can'st thou bind the Unicorn with his Band to labour in the Furrow, or will he plough the Valleys after thee?*'"

Topsell dilates at great length on the Unicorn. He agrees with Spenser and Guillim, and says:—" These Beasts are very swift, and their legges have no Articles (*joints*). They keep for the most part in the desarts, and live solitary in the tops of the Mountaines. There was nothing more horrible than the voice or braying of it, for the voice is strain'd above measure. It fighteth both with the mouth and with the heeles, with the mouth biting like a Lyon, and with the heeles kicking like a Horse. . . . He feereth not Iron nor any yron Instrument (as *Isodorus* writeth) and that which is most strange of all other, it fighteth with his owne kind, yea even with the females unto death, except when it burneth in lust for procreation: but unto straunger Beasts, with whome he hath no affinity in nature, he is more sotiable and familiar, delighting in their company when they come willing unto him, never rising against them; but, proud of their dependence and retinue, keepeth with them all quarters of league and truce; but with his female, when once his flesh is tickled with lust, he groweth tame, gregall, and loving, and so continueth till she is filled and great with young, and then returneth to his former hostility."

There was a curious legend of the Unicorn, that it would, by its keen scent, find out a maiden, and run to her, laying its head in her lap. This is often used as an emblem of the Virgin Mary, to denote her purity. The following is from the Bestiary of Philip de Thaun, and, as its old French is easily read, I have not translated it :—

> "Monoceros est Beste, un corne ad en la teste,
> Purceo ad si a nun, de buc ad façun ;
> Par Pucele est prise ; or vez en quel guize.
>   Quant hom le volt cacer et prendre et enginner,
> Si vent hom al forest ù sis riparis est ;
> Là met une Pucele hors de sein sa mamele,
> Et par odurement Monosceros la sent ;
> Dunc vent à la Pucele, et si baiset la mamele,
> En sein devant se dort, issi veut à sa mort ;
> Li hom suivent atant ki l'ocit en dormant
> U trestont vif le prent, si fais puis sun talent.
>   Grant chose signifie." . . .

Topsell, of course, tells the story :—"It is sayd that Unicorns above all other creatures, doe reverence Virgines and young Maides, and that many times at the sight of them they grow tame, and come and sleepe beside them, for there is in their nature a certaine savor, wherewithall the Unicornes are allured and delighted ; for which occasion the *Indian* and *Ethiopian* hunters use this stratagem to take the beast. They take a goodly, strong, and beautifull young man, whom they dresse in the Apparell of a woman, besetting him with divers odoriferous flowers and spices.

"The man so adorned they set in the Mountaines or Woods, where the Unicorne hunteth, so as the wind may carrie the savor to the beast, and in the meane season the other hunters hide themselves : the Unicorne deceaved with the outward shape of a woman, and sweete smells, cometh to the young man without feare, and so suffereth his head to bee covered and wrapped within his large sleeves, never stirring, but lying still and asleepe, as in his most acceptable repose. Then, when the hunters, by the signe of the young man, perceave him fast and secure, they come uppon him, and, by force, cut off his horne, and send him away alive: but, concerning this opinion wee have no elder authoritie than *Tzetzes*, who did not live above five hundred yeares agoe, and therefore I leave the reader to the freedome of his owne judgment, to believe or refuse this relation; neither is it fit that I should omit it, seeing that all writers, since the time of *Tzetzes*, doe most constantly beleeve it.

"It is sayd by *Ælianus* and *Albertus*, that, except they bee taken before they bee two yeares old they will never bee tamed; and that the Thrasians doe yeerely take some of their Colts, and bring them to their King, which he keepeth for combat, and to fight with one another; for when they are old, they differ nothing at all from the most barbarous, bloodie, and ravenous beasts. Their flesh is not good for meate, but is bitter and unnourishable."

It is hardly worth while to go into all the authorities treating of the Unicorn; suffice it to say, that it was an universal belief that there were such animals in existence, for were not their horns in proof thereof? and were they not royal presents fit for the mightiest of potentates to send as loving pledges one to another? for it was one of the most potent of medicines, and a sure antidote to poison. And they were very valuable, too, for Paul Hentzner—who wrote in the time of Queen Elizabeth—says that, at Windsor Castle, he was shown, among other things, the horn of an Unicorn of above eight spans and a half in length, *i.e.*, about $6\frac{1}{2}$ feet, valued at £10,000. Considering that money was worth then about three times what it is now, an Unicorn's horn was a right royal gift.

Topsell, from whom I have quoted so much, is especially voluminous and erudite on Unicorns; indeed, in no other old or new author whom I have consulted are there so many facts (?) respecting this fabled beast to be found. Here is his history of those horns then to be found in Europe :—

"There are two of these at *Venice* in the Treasurie of S. *Marke's* Church, as *Brasavolus* writeth, one at *Argentoratum*, which is wreathed about with divers sphires.[1] There are also two in the Treasurie of the King of *Polonia*, all of them as long as a man in his stature. In the yeare 1520, there was found the horne of a *Unicorne* in the river *Arrula*, neare *Bruga* in Helvetia, the upper face or out side whereof was a darke yellow; it was two cubites (3 *feet*) in length, but had upon it no plights[2] or wreathing versuus. It was very odoriferous (especially when any part of it was set on fire), so that it smelt like muske: as soone as it was found, it was carried to a Nunnery called *Campus regius*, but, afterwards by the Governor of *Helvetia*, it was recovered back againe, because it was found within his teritorie. . . .

"Another certaine friend of mine, being a man worthy to be beleeved, declared unto me that he saw at *Paris*, with the Chancellor, being Lord of *Pratus*, a peece of a Unicorn's horn, to the quantity of a cubit, wreathed in tops or spires, about the thicknesse of an indifferent staffe (the compasse therof extending to the quantity of six fingers) being within, and without, of a muddy colour, with a solide substance, the fragments whereof would boile in the Wine although they were never burned, having very little or no smell at all therein.

---

[1] Spirals.   [2] Plaits.

"When *Joannes Ferrerius* of *Piemont* had read these thinges, he wrote unto me, that, in the Temple of *Dennis*, neare unto *Paris*, that there was a Unicorne's horne six foot long, . . . but that in bignesse, it excceded the horne at the Citty of *Argentorate*, being also holow almost a foot from that part which sticketh unto the forehead of the Beast, this he saw himselfe in the Temple of S. *Dennis*, and handled the horne with his handes as long as he would. I heare that in the former yeare (which was from the yeare of our Lord), 1553, when *Vercella* was overthrown by the French, there was broght from that treasure unto the King of France, a very great Unicorn's horne, the price wherof was valued at fourscore thousand Duckets.[1]

"*Paulus Powius* describeth an Unicorne in this manner; That he is a beast, in shape much like a young Horse, of a dusty colour, with a maned necke, a hayry beard, and a forehead armed with a Horne of the quantity of two Cubits, being seperated with pale tops or spires, which is reported by the smoothnes and yvorie whitenesse thereof, to have the wonderfull power of dissolving and speedy expelling of all venome or poison whatsoever.

"For his horne being put into the water, driveth away the poison, that he may drinke without harme, if any venemous beast shall drinke therein before him. This cannot be taken from the Beast, being alive, for as much as he cannot possible be taken by any deceit: yet it is usually seene that the horne is found in the desarts, as it happeneth in Harts, who cast off their olde horne

---

[1] Taking the Ducat at 9s. 4½d., it would come to £37,000, but if this were multiplied by three, the lowest computation of the value of money then, and now, it would be worth considerably over £100,000.

thorough the inconveniences of old age, which they leave unto the Hunters, Nature renewing an other unto them.

"The horne of this beast being put upon the Table of Kinges, and set amongest their junkets and bankets, doeth bewray the venome, if there be any suche therein, by a certaine sweat which commeth over it. Concerning these hornes, there were two seene, which were two cubits in length, of the thicknesse of a man's Arme, the first at *Venice*, which the Senate afterwards sent for a gift unto *Solyman* the Turkish Emperor: the other being almost of the same quantity, and placed in a Sylver piller, with a shorte or cutted[1] point, which *Clement* the Pope or Bishop of *Rome*, being come unto *Marssels* brought unto *Francis* the King, for an excellent gift." . . . They adulterated the real article, for sale. "*Petrus Bellonius* writeth, that he knewe the tooth of some certaine Beast, in time past, sold for the horne of a Unicorne (what beast may be signified by this speech I know not, neither any of the French men which do live amongst us) and so smal a peece of the same, being adulterated, sold 'sometimes for 300 Duckets.' But, if the horne shall be true and not counterfait, it doth, notwithstanding, seeme to be of that creature which the Auncientes called by the name of an Unicorne, especially *Ælianus*, who only ascribeth to the same this wonderfull force against poyson and most grievous diseases, for he maketh not this horne white as ours doth seeme, but outwardly red, inwardly white, and in the Middest or secretest part only blacke."

Having dilated so long upon the Unicorn, it would be a pity not to give some idea of the curative properties of

---

[1] Another name for short—vide *Cutty pipe*—*Cutty sark*.

its horn—always supposing that it could be obtained genuine, for there were horrid suspicions abroad that it might be "the horne of some other beast brent in the fire, some certaine sweet odors being thereunto added, and also imbrued in some delicious and aromaticall perfume. Peradventure also, Bay by this means, first burned, and afterwards quenched, or put out with certaine sweet smelling liquors." To be of the proper efficacy it should be taken new, but its power was best shown in testing poisons, when it sweated, as did also a stone called "the Serpent's tongue." And the proper way to try whether it was genuine or not, was to give Red Arsenic or Orpiment to two pigeons, and then to let them drink of two samples; if genuine, no harm would result—if adulterated, or false, the pigeons would die.

It was also considered a cure for Epilepsy, the Pestilent Fever or Plague, Hydrophobia, Worms in the intestines, Drunkenness, &c., &c.,—and it also made the teeth clean and white;—in fact, it had so many virtues that "no home should be without it."

And all this about a Narwhal's horn!

The Rhinoceros.

The true Unicorn is, of course, the Rhinoceros, and this picture of it is as early an one as I can find, being taken from Aldrovandus de Quad, A.D. 1521. Gesner and Topsell both reproduce it, at later dates, but *reversed*. The latter says that Gesner drew it from the life at Lisbon—but having Aldrovandus and the others before me, I am bound to give the palm to the former,

and confess the others to be piracies. It is certain, however, that whoever drew this picture of a Rhinoceros must have seen one, either living or stuffed, for it is not too bizarre.

Topsell approaches this animal with an awe and reverence, such as he never shows towards any other beast; indeed, he gets quite solemn over it, and he thus commences his *Apologia:*—" But for my part, which write the English story, I acknowledge that no man must looke for that at my hands, which I have not received from some other: for I would bee unwilling to write anything untrue, or uncertaine out of mine owne invention; and truth on every part is so deare unto mee, that I will not lie to bring any man in love and admiration with God and his works, for God needeth not the lies of men : To conclude, therefore, this Præface, as the beast is strange, and never seene in our countrey, so my eyesight cannot adde anything to the description; therefore harken unto that which I have observed out of other writers."

They were very rare beasts, among the early Roman Emperors, but in the later Empire they were introduced into the Circus, but many centuries rolled on before we, in England, were favoured with a sight of this great animal. Topsell had not seen one, and he wrote in 1607, so we accept his *Apologia* with all his errors :—
" *Oppianus* saith that there was never yet any distinction of sexes in these *Rhinocerotes;* for all that ever have been found were males, and not females, but from hence let no body gather that there are no females, for it were impossible that the breede should continue without females.

" When they are to fight they whet their horne upon a

stone, and there is not only a discord between these beasts and Elephants for their food, but a natural description and enmity: for it is confidently affirmed, that when the Rhinoceros which was at *Lisborne*, was brought into the presence of an Elephant, the Elephant ran away from him. How and what place he overcometh the Elephant, we have shewed already in his story, namely, how he fastneth his horne in the soft part of the Elephantes belly. He is taken by the same meanes that the *Unicorne* is taken, for it is said by *Albertus, Isodorus*, and *Alumnus*, that above all other creatures they love Virgins, and that unto them they will come be they never so wilde, and fall a sleepe before them, so being asleepe they are easily taken, and carried away. All the later Physitians do attribute the vertue of the *Unicorn's* horne to the *Rhinocereos* horn."

Ser Marco Polo, speaking of Sumatra, or, as he called it, Java the Less, says in that island there are numerous unicorns. "They have hair like that of a buffalo, feet like those of an elephant, and a horn in the middle of the forehead, which is black and very thick. They do no mischief, however, with the horn, but with the tongue alone; for this is covered all over with long and strong prickles, (and when savage with any one they crush him under their knees, and then rasp him with their tongue). The head resembles that of a wild boar, and they carry it ever bent towards the ground. They delight much to abide in mire and mud. 'Tis a passing ugly beast to look upon, and is not in the least like that which our stories tell us of as being caught in the lap of a virgin; in fact, 'tis altogether different from what we fancied."

The Gulo.

Olaus Magnus thus describes the Gulo or Gulon:—
"Amongst all creatures that are thought to be insatiable in the Northern parts of *Sweden*, the *Gulo* hath his name to be the principall; and in the vulgar tongue they call him *Jerff*, but in the *German* language *Vielfras*; in the Sclavonish speech *Rossamaka*, from his much eating, and the Latin name is *Gulo*, for he is so called from his gluttony. He is as great as a great Dog, and his ears

and face are like a Cat's: his feet and nails are very sharp; his body is hairy, with long brown hair, his tail is like the Foxes, but somewhat shorter, but his hair is thicker, and of this they make brave Winter Caps. Wherefore this Creature is the most voracious; for, when he finds a carcasse, he devours so much, that his body, by over-much meat, is stretched like a Drum, and finding a streight (*narrow*) passage between Trees, he presseth between them, that he may discharge his body by violence; and being thus emptied, he returns to the carcasse, and fills himself top full; and then he presseth again through the same narrow passage, and goes back to the carkasse, till he hath devoured it all; and then he hunts eagerly for another. It is supposed he was created by nature to make men blush, who eat and drink till they spew, and then feed again, eating day and night, as *Mechovita* thinks in his *Sarmatia*. The flesh of this Creature is altogether uselesse for man's food; but his skin is very commodious and pretious. For it is of a white brown black colour, like a damask cloth wrought with many figures; and it shews the more beautiful, as by the Industry of the Artist it is joyn'd with other garments in the likenesse or colour. Princes and great men use this habit in Winter, made like Coats; because it quickly breeds heat, and holds it long; and that not onely in *Swethland*, and *Gothland*, but in *Germany*, where the rarity of these skins makes them to be more esteemed, when it is prised in ships among other Merchandise.

"The Inhabitants are not content to let these skins be transported into other Countries, because, in Winter, they use to entertain their more noble guests in these skins; which is a sufficient argument that they think nothing more comely and glorious, than to magnifie at all times, and in all orders their good guests, and that in the most vehement cold, when amongst other good turns they cover their beds with these skins.

"And I do not think fit to overpasse, that when men sleep under these skins, they have dreams that agree with the nature of that Creature, and have an insatiable stomach, and lay snares for other Creatures, and prevent them themselves. It may be that it is as they that eat hot

Spices, Ginger or Pepper seem to be inflamed; and they that eat Sugar seem to be choked in water. There seems to be another secret of Nature in it, that those who are clothed in those Skins, seem never to be satisfied.

"The guts of this Creatures are made into strings for Musicians, and give a harsh sound, which the Natives take pleasure in; but these, tempered with sweet sounding strings, will make very good Musick. Their hoofs made like Circles, and set upon heads subject to the Vertigo, and ringing ears, soon cure them. The Hunters drink the blood of this beast mingled with hot water; also seasoned with the best Honey, it is drunk at Marriages. The fat, or tallow of it, smeered on putrid Ulcers for an ointment is a sudden cure. Charmers use the teeth of it. The hoofs, newly taken off, will drive away Cats and Dogs, if they do but see it, as birds fly away, if they spy but the Vultur or the Bustard.

"By the Hunter's various Art, this Creature is taken onely in regard of his pretious skin; and the way is this;—They carry into the wood a fresh Carkasse; where these beasts are wont to be most commonly; especially in the deep snows (for in Summer their skins are nothing worth) when he smels this he falls upon it, and eats till he is forced to crush his belly close between narrow trees, which is not without pain; the Hunter, in the mean time, shoots, and kills him with an arrow.

"There is another way to catch this Beast, for they set Trees, bound asunder with small cords, and these fly up when they eat the Carkasse, and strangle them; or else he is taken, falling into pits dug upon one side, if the Carkasse be cast in, and he is compelled by hunger to feed upon it. And there is hardly any other way to catch him with dogs, since his claws are so sharp, that dogs dare not encounter with him, that fear not to set upon the most fierce Wolves."

Of this animal Topsell says :—" This beast was not known by the ancients, but hath bin since discovered in the Northern parts of the world, and because of the great voracity thereof, it is called *Gulo*, that is, a devourer ; in imitation of the Germans, who call such devouring Creatures *Vilsruff*, and the Swedians *Cerff*, and in *Lituania* and *Muscovia* it is called *Rossomokal*. It is thought to be engendered by a *Hyæna* and a *Lionesse*, for in quality it resembleth a *Hyæna*, and it is the same which is called *Crocuta :* it is a devouring and unprofitable creature having sharper teeth than other creatures. Some thinke it is derived from a wolf and a dog, for it is about the bignesse of a dog. It hath the face of a Cat, the body and taile of a Foxe ; being black of colour ; his feet and nailes be most sharp, his skin rusty, the haire very sharp, and it feedeth upon dead carkases."

He then describes its manner of feeding, evidently almost literally copying Olaus Magnus, and thus continues :—" There are of these beastes two kindes, distinguished by coulour, one blacke, and the other like a Wolfe : they seldom kill a man or any live beastes, but feede upon carrion and dead carkasses, as is before saide, yet, sometimes, when they are hungry, they prey upon beastes, as horses and such like, and then they subtlely ascend up into a tree, and when they see a beast under the same, they leape downe upon him and destroy him. A Beare is afraide to meete them, and unable to match them, by reason of their sharpe teeth.

" This beast is tamed, and nourished, in the courts of Princes, for no other cause than for an example of incredible voracitie. When he hath filled his belly, if he can find no trees growing so neare another, as by sliding betwixte them, hee may expell his excrements, then taketh he an Alder-tree, and with his forefeete rendeth the same asunder, and passeth through the middest of it, for the cause aforesaid. When they are wilde, men kill them with bowes and guns, for no other cause than for their skins, which are pretious and profitable, for they are white spotted, changeably interlined like divers flowers, for which cause the greatest princes, and richest nobles use them in garments in the Winter time; such are the Kings of *Polonia, Swede-land, Goat-land,* and the princes of *Germany.* Neither is there any skinne which will sooner take a colour, or more constantly retaine it. The outward appearance of the saide skinne is like to a damaskt garment, and besides this outward parte there is no other memorable thing woorthy observation in this ravenous beast, and therefore, in *Germany,* it is called a foure-footed Vulture."

As a matter of fact, the Glutton or Wolverine, which is not unlike a small bear, can consume (while in confinement) thirteen pounds of meat in a day. In its wild state, if the animal it has killed is too large for present consumption, it carries away the surplus, and stores it up in a secure hiding-place, for future eating.

The Bear.

As Pliny not only uses all Aristotle's matter anent Bears, but puts it in a consecutive, and more readable form, it is better to transcribe his version than that of the older author.

" Bears couple in the beginning of winter. The female then retires by herself to a separate den, and then brings forth, on the thirtieth day, mostly five young ones. When first born, they are shapeless masses of white flesh, a little larger than mice; their claws alone being prominent. The mother then licks them into proper shape.[1] The male remains in his retreat for forty days, the female four months. If they happen to have no den, they construct a retreat with branches and shrubs, which is made impenetrable to the rain, and is lined with soft leaves. During the first fourteen days they are overcome by so deep a sleep, that they cannot be aroused by wounds even. They become wonderfully fat, too, while in this lethargic state. This fat is much used in medicine, and it is very useful in preventing the hair from falling off.[2] At the end of these fourteen days they sit up, and find nourishment by sucking their fore paws. They warm their cubs, when cold, by pressing them to the breast, not unlike the way in which birds brood over their eggs. It is a very astonishing thing, but Theophrastus believes it, that if we preserve the flesh of the bear, the animal being killed in its dormant state, it will increase in bulk, even though it may have been cooked. During this period no signs of food are to be found in the stomach

[1] "An unlicked cub" is a proverb which has sprung from this fable. Aristotle was right when he said that bears when newly born were without hair, and blind, but wrong in continuing "its legs, and almost all its parts, are without joints." Still, the popular idea that bears licked their young into shape, lasted till very modern times, and still survives in the proverb quoted. Shakespeare mentions it in 3 Henry VI. iii. 2:—

" Like to Chaos, or an unlick'd bear whelp,
That carries no impression like the dam."

And Chester, in his *Love's Martyr*, speaking of the Bear, says—

" Brings forth at first a thing that's indigest,
A lump of flesh without all fashion,
Which she, by often licking brings to rest,
Making a formal body, good and sound.
Which often in this iland we have found."

[2] This use of bear's grease is about 1800 years old.

of the animal, and only a very slight quantity of liquid; there are a few drops of blood only, near the heart, but none whatever in any other part of the body. They leave their retreat in the spring, the males being remarkably fat; of this circumstance, however, we cannot give any satisfactory explanation, for the sleep, during which they increase so much in bulk, lasts, as we have already stated, only fourteen days. When they come out, they eat a certain plant, which is known as *Aros*, in order to relax the bowels, which would otherwise become in a state of constipation; and they sharpen the edges of their teeth against the young shoots of the trees.

"Their eyesight is dull, for which reason in especial, they seek the combs of bees, in order that from the bees stinging them in the throat, and drawing blood, the oppression in the head may be relieved. The head of the bear is extremely weak, whereas, in the lion, it is remarkable for its strength: on which account it is, that when the bear, impelled by any alarm, is about to precipitate itself from a rock, it covers its head with its paws. In the arena of the Circus they are often to be seen killed by a blow on the head with the fist. The people of Spain have a belief, that there is some kind of magical poison in the brain of the bear, and therefore burn the heads of those that have keen killed in their public games; for it is averred, that the brain, when mixed with drink, produces, in man, the rage of the bear.

"These animals walk on two feet, and climb trees backwards. They can overcome the bull, by suspending themselves, by all four legs, from his muzzle and horns, thus wearing out its powers by their weight. In no other animal is stupidity found more adroit in devising mischief."

Olaus Magnus, in writing about bears, gives precedence to the white, or Arctic bear, and gives an insight into the religious life of the old Norsemen, who, when converted, thought their most precious things none too good for the "Church." If we consider the risk run in obtaining a white bear's skin, and the privations and cold endured in getting it, we may look upon it as a Norse treasure. "Silver and Gold have I none; but such as I have, give I unto thee." He gives a short, but truthful account of their habits, and winds up his all too brief narration thus :—"These white Bear Skins are wont to be offered by the Hunters, for the high Altars of Cathedrals, or Parochial Churches, that the Priest celebrating Mass standing, may not take cold of his feet, when the Weather is extream cold. In the Church at *Nidrosum*, which is the Metropolis of the Kingdom of *Norway*, every year such white Skins are found, that are faithfully offered by the Hunters Devotion, whensoever they take them, and Wolves-Skins to buy Wax-Lights, and to burn them in honour of the Saints."

Olaus Magnus is very veracious in his dealings with White Bears, but he morally retrogrades when he touches upon the Black and Brown Bears. The illustrations of this portion of Olaus Magnus are exceedingly graphic. In treating of the cunning used in killing bears, he says :—" In killing black and cruel Bears in the Northern Kingdoms, they use this way, namely, that when, in Autumn the Bear feeds on certain red ripe Fruit (*Query Cranberries*) on trees that grow in Clusters like Grapes, either going up into the Trees, or standing on the ground, and pulling down the Trees, the cunning Hunter, with broad Arrows from a Crosse-bow shoots at him, and these pierce deep; and he is so suddenly moved with this fright, and wound received, that he presently voids backward all the Fruit he ate, as Hailstones; and presently runs upon an Image of a man made of wood, that is set purposely before him, and rends and tears

that, till another Arrow hit him, that gives him his death's wound, shot by the Hunter that hides himself behind some Stone or Tree. For when he hath a wound, he runs furiously, at the sight of his blood, against all things in his way, and especially the Shee-Bear, when she suckleth her Whelps.

"The Bears watch diligently for the passing of Deer; and chiefly, the Shee-Bear when she hath brought forth her Whelps; who not so much for Hunger, as for fearing of losing her Whelps, is wont to fall cruelly upon all she meets. For, she being provoked by any violence, far exceeds the force of the He-Bear, and Craft, that she may revenge the loss of her Young. For she lyes hid amongst the thick boughs of Trees, and young

Shoots; and if a Deer, trusting to the glory of his horns, or quick smell, or swift running, come too neare that place unawares, she suddenly falls out upon him to kill him; and if he first defend himself with his horns, yet he is so tired with the knots and weight of them, being driven by the rage of the Bear, that he is beaten to the ground, that losing force and life, he falls down a prey to be devoured. Then she will set upon the Bull with his horns, using the same subtilty, and casts

herself upon his back; and when the Bull strives with his horns to cast off the Bear, and to defend himself, she fasteneth on his horns and shoulders with her paws, till, weary of the weight he falls down dead. Then laying the Bull on his back like a Wallet, she goes on two feet into the secret places of the Woods to feed upon him. But when, in Winter she is hunted, she is betrayed by Dogs, or by the prints of her feet in the Snow, and can hardly escape from the Hunters that run about her from all sides."

Magnus then retails the usual fables about bears

licking their young into shape, their building houses, &c., &c., after which he discourses about the bear and hedgehog, a story which has nothing to do with the picture. It is described as "the Battail between the Hedge-Hog, and the Bear."

"Though the *Urchin* have sharp pointed prickles, whereby he gathereth Apples to feed on, and these he

hides in hollow Trees, molesting the *Bear* in his Den: yet is he oppressed by the cunning and weight of the *Bear:* namely when the Urchin roles himself up round as a ball, that there is nothing but his prickles to come at: yet with this means he cannot prevail against the *Bear*, which opens him, to revenge the wrong he did her in violating her Lodging. Nor can the *Bear* eat the *Hedge-Hog*, it is such miserable poor and prickly meat. Wherefore returning again into his Cave, he sleeps, and grows fat, living by sucking his paw.

"The *Bears* also fight against the *Bores*, but seldome get the victory, because they can better defend themselves with their Tusks, than the *Bull* or the *Deer* can by their

Horns, or running swiftly. The strong *Horses* keep off the *Bears* with their biting and kicking, from the *Mares* that are great with *Foals*. Young *Colts* save themselves by running, but they will always hold this fear, and so become unprofitable for the Wars. Wherefore they use this stratagem: some Souldier puts on a Bear's skin, and meets them, by reason that they are horses that the Bears have hunted."

The Northern Bears seem to have been wonderful creatures, for they used to go mad after eating Mandragora, and then they were in the habit of making a meal off ants, by way of recovering their sanity. They were then, as now, noted for their love of honey, and this

illustration depicts them as coming out of, and going into the ground after bees and honey; nay, it would seem as if they even invaded the barrels put up in the trees to serve as hives. But man was more cunning than they, and a good bear-skin in those cold regions, had a value far exceeding honey.

"Since that in the Northern Countries, especially *Podolia, Russia*, and places adjacent, because of the great multitude of Bees, the Hives at home will not contain them, the Inhabitants willingly let them fly unto hollow Trees, made so by Nature, or by Art, that they may increase there. Wherefore mortal stratagems are thus prepared for Bears, that use to steal honey (for they having a most weak head, as a Lion hath the strongest, for sometimes they will be killed with a blow under their ear); namely a Woodden Club set round with Iron

points is hung over the hole the Bees come forth of, from some high bough, or otherwise; and this, being cast upon the head of the greedy Bear that is going to steal the honey, kills him striving against it; so he loseth his life, flesh, and skin to the Master, for a little honey. Their flesh is salted up like Hog's flesh, Stag's flesh, Elk's, or Ranged deer's flesh, to eat in Camps, and the Tallow of them is good to cure any wounds."

Every one of my readers, who is not a Scotsman, will appreciate the delicate musical taste of the bear, in the matter of bagpipes—Bruin cannot stand the skirling, and, in the illustration, seems to be remonstrating with the piper.

"It is well enough known that Bears, Dolphins, Stags, Sheep, Calves and Lambs, are much delighted with Musick : and, again, they are to be driven from their Heards by some harsh sounding Pipes, or Horns, that when they hear the sound they will be gone into the Woods, a great way off. Now the Shepheards of the Cattel know this well enough : they will play upon their two horned Pipes continually, which sometimes are taken away by Bears, until such time as the Bear is forced by Hunger to go away to get his food. Wherefore they

take a Goat's Horn, and sometimes a Cow's Horn, and make such a horrid noise, that they scare the wild beasts, and so return safe to their dispersed flocks. This two horned Pipe, which in their tongue they call *Seec-Pipe*, they carry to the fields with them, for they have learned by use, that their Flocks and Heards will feed the better and closer together.

"The *Russians* and *Lithuanians* are more near to the Swedes and Goths on the Eastern parts : and these hold it a singular delight, to have always the most cruel

Beasts bred up tame with them, and made obedient to their commands in all things. Wherefore to do this the Sooner, they keep them in Caves, or tyed with Chains, chiefly Bears newly taken in the Woods, and half starve them; and they appoint one or two Masters, cloathed one like the other, to carry Victuals to them, that they may be accustomed to play with them, and handle them when they are loose. Also they play on Pipes sweetly, and with this they are much taken: and thus they use them to sport and dance, and then, when the Pipes sound differently, they are taught to lift up their legs, as by a more sharp sign, to end the Dance with, that they may go on their hinder feet, with a Cap in their fore feet, held out to the Women and Maids, and others that saw them dance, and ask a reward for their dancing; and, if it is not given freely, they will murmure, as they are directed by their Master, and will nod their heads, as desiring them to give more money: So the Master of these Bears, that cannot speak the language of other countries, will get a good gain by his dumb Beast. Nor doth this seem to be done onely because that these should live by this small gain; for the Bearherds that lead these Bears, are, at least, ten or twelve lusty men; and in their company, sometimes, there go Noblemen's sons, that they may learn the manners, fashions, and distances of places, the Military Arts, and Concord of Princes, by these merry Pastimes. But since they were found, in *Germany*, to spoil Travellers, and to cast them to their Bears to eat, most strict Laws are made against them, that they may never come there again.

"There is another Sport, when Bears taken, are put into a Ship, and shew merry pastimes in going up and down the Ropes, and sometimes are profitable for some unexpected accident. For Histories of the Provincials mention, that it hapned, that one was thus freed from a Pirate that was like to set upon him; for the Pirate coming on, was frighted at it, when he saw afar off, men, as he supposed, going up and down the Ropes, from the Top Mast, as the manner is to defend the Ship. Whereas they were but young Bears, playing on the Ropes. But the most pleasant sight of all is, that when the Bears look out of the Ship into the

Waters, a great number of Sea Calves will come and gaze upon them, that you would think an innumerable Company of Hogs swam about the Ship, and they are caught by the Sea men with long Spears, with Hooks, and a Cord tyed to them; and so are also the other Beasts, that come to help the Sea Calves, taken, and crying like to Hogs. Also the Bears are let down to swim, that they may catch these wandering Sea-Calves, or else, when it thunders, and the weather is tempestuous, they be taken above Water.

"But that tame Bears may not onely be kept unprofitably to feed, and make sport, they are set to the Wheels in the Courts of great men, that they may draw up Water out of deep Wells; and that in huge Vessels made for this purpose, and they do not help alone this Way, but they are set to draw great Waggons, for they are very strong in their Legs, Claws, and Loins; nor is it unfit to make them go upright, and carry burdens of Wood, and such like, to the place appointed, or they stand at great men's doors, to keep out other hurtful Creatures. When they are young, they will play wonderfully with Boys, and do them no hurt."

Topsell goes through the usual stories of bears licking their cubs into shape, and subsisting by sucking their claws—but he also affords us much information about bears, which we do not find in modern Natural Histories:—"At what time they come abroad, being in the beginning of May, which is the third moneth from the Spring. The old ones being almost dazled with long darknes, comming into light againe, seeme to stagger and reele too and fro, and then for the straightnesse of their guts, by reason of their long fasting, doe eat the herbe *Arum*, called in English *Wake-Robbin*, or *Calves-foot*, being of very sharpe and tart taste, which enlargeth their guts, and so, being recovered, they remaine all the time their young are with them, more fierce, and cruell than at other times. And concerning the same *Arum*, called also *Dracunculus*, and *Oryx*, there is a pleasant vulgar tale, whereby some have conceived that Beares eat this herbe before their lying secret, and by vertue thereof (without meat, or sence of cold) they passe away the whole winter in sleepe.

"There was a certaine cow-heard, in the Mountains

of *Helvetia*, which, comming downe a hill, with a great caldron on his backe, he saw a beare eating a root which he had pulled up with his feet; the cowheard stood still till the beare was gone, and afterward came to the place where the beast had eaten the same, and, finding more of the same roote, did likewise eat it; he had no sooner tasted thereof, but he had such a desire to sleepe, that hee could not containe himselfe, but he must needs lie down in the way, and there fell a sleep, having covered his heade with the caldron, to keep himself from the vehemency of the colde, and there slept all the Winter time without harme, and never rose againe till the spring time; which fable if a man will beleeve, then, doubtlesse, this hearbe may cause the Beares to be sleepers, not for fourteene dayes, but for fourscore dayes together.

"The ordinary food of Beares is fish; for the Water beare, and others will eate fruites, Apples, Grapes, Leaves, and Pease, and will breake into bee hives sucking out the honey; likewise Bees, Snayles and Emmets, and flesh, if it bee leane, or ready to putrifie; but, if a Beare doe chance to kill a swine, or a Bull, or Sheepe, he eateth them presentlie, whereas other beasts eate not hearbes, if they eate flesh : likewise they drinke water, but not like other beastes, neither sucking it, or lapping it, but as it were, even bitinge at it.

"They are exceeding full of fat or Larde-greace, which some use superstitiouslie beaten with oile, wherewith they anoint their grape-sickles when they go to vintage, perswading themselves that if no bodie knows thereof, their tender vine braunches shall never be consumed by catterpillers.

"Others attribute this to the vertue of Beare's blood, and *Theophrastus* affirmeth, that if beare's grease be kept in a vessell, at such time as the beares lie secret, it will either fill it up, or cause it to runne over. The flesh of beares is unfit for meate, yet some use to eate it, after it hath been twice sodden; other eat it baked in pasties, but the truth is, it is better for medicine than food. *Theophrastus* likewise affirmeth, that at the time when beares lie secret, their dead flesh encreaseth, which is kept in houses, but beare's fore feet are held for a verie delicate and well tasted foode, full of sweetnes, and much used by the German Princes.

"And because of the fiercenesse of this beast, they are seldome taken alive, except they be very young, so that some are killed in the Mountaines by Poyson, the Country being so steepe and rocky that hunters cannot followe them; some taken in ditches of the earth and other ginnes. *Oppianus* relateth that neare *Tygris* and *Armenia*, the inhabitauntes use this Stratigem to take Beares.

"The people go often to the Wooddes to find the Denne of the Beare, following a leam hound, whose nature is, so soone as he windeth the beast, to barke, whereby his leader discovereth the prey, and so draweth off the hounde with the leame; then come the people in great multitude, and compasse him about with long nets, placing certaine men at each end: then tie they a long rope to one side of the net, as high from the ground, as the small of a Man's belly; whereunto are fastned divers plumes and feathers of vultures, swannes, and other resplendant coloured birdes, which, with the wind make a noise or hissing, turning over and glistering; on the other side of the net they build foure little hovels of greene boughes, wherein they lay foure men covered all over with greene leaves; then, all being prepared, they sound their Trumpets, and wind their horns; at the noise whereof the beare ariseth, and in his fearefull rage runneth too and fro as if he sawe fire : the young men, armed, make unto him, the beare, looking round about, taketh the plainest way toward the rope hung full of feathers, which, being stirred, and haled by those that holde it, maketh the beare much affraid with the ratling and hissing thereof, and so flying from that side halfe mad, runneth into the nets, where the keepers entrap him so cunningly, that he seldome escapeth.

"When a Beare is set upon by an armed man, he standeth upright, and taketh the man betwixt his forefeet, but he, being covered all over with yron plates can receive no harm, and then may easily, with a sharpe knife or dagger pierce thorough the heart of the beast.

"If a shee beare having young ones be hunted, shee driveth her Whelpes before her, untill they be wearied, and then, if she be not prevented, she climbeth uppon a tree, carrying one of her young in her mouth, and the other on her backe. A Beare will not willingly fight with a man, but, being hurt by a man, he gnasheth his teeth, and licketh his forefeete, and it is reported by an Ambassador of *Poland*, that when the *Sarmatians* finde a beare, they inclose the whole Wood by a multitude of people standing not above a cubit one from another ; then cut they downe the outmost trees, so that they raise a Wall of wood to hemme in the Beares ; this being effected, they raise the Beare, having certaine forkes in their hands, made for that purpose, and, when the Beare approacheth, they, (with those forkes) fall upon him, one keeping his head, another one leg, other his body, and so, with force, muzzle him and tie his legges, leading him away. The Rhætians use this policy to take Wolves and Beares ; they raise up great posts, and crosse them with a long beame laded with heavy weightes, unto the which beame they fasten a corde with meat therein, whereunto the beast comming, and biting at the meat, pulleth downe the beame upon her owne pate.

"The inhabitants of *Helvetia* hunt them with mastiffe Dogges, because they should not kill their cattell left at large in the fielde in the day time; They likewise shoote them with gunnes, giving a good summe of money to them that can bring them a slaine beare. The *Sarmatians* use to take Beares by this sleight; under those trees wherein bees breed, they plant a great many of sharpe pointed stakes, putting one hard into the hole wherein the bees go in and out, whereunto the Beare climbing, and comming to pull it forth, to the end that she may come to the hony, and being angry that the stake sticketh so fast in the hole, with violence plucketh it foorth with both her fore feet, whereby she looseth her holde, and falleth downe upon the picked stakes, whereupon she dieth, if they that watch for her come not to take her off. There was reported by *Demetrius*, Ambassador at *Rome*, from the King of *Musco*, that a neighbor of his, going to seek hony, fell into a hollow tree, up to the brest in hony, where he lay two days, being not heard by any man to complain; at length came a great Beare to this hony, and, putting his head into the tree, the poore man tooke hold thereof, whereat, the Beare, suddenly affrighted, drew the man out of that deadly danger, and so ranne away for feare of a worse creature.

"But, if there be no tree wherein Bees doe breed neere to the place where the Beare abideth, then they use to annoint some hollow place of a tree with hony, whereinto Bees will enter and make hony combes, and when the Beare findeth them, she is killed as aforesaide. In

Norway they use to saw the tree almost asunder, so that when the beast climbeth it, she falleth downe upon piked stakes laid underneath to kill her; and some make a hollow place in a tree, wherein they put a great pot of water, having annointed it with hony, at the bottome wherof are fastened certaine hookes bending downeward, leaving an easie passage for the beare to thrust in her head to get the honie, but impossible to pull it foorth againe alone, because the hookes take holde on her skinne; this pot they binde fast to a tree, whereby the Beare is taken alive and blinde folded, and though her strength breake the corde or chaine wherewith the pot is fastened, yet can shee not escape or hurt any bodie in the taking, by reason her head is fastened in the pot.

" To conclude, other make ditches or pits under Apple trees, laying upon their mouth rotten stickes, which they cover with earth, and strawe uppon it herbes, and when the beare commeth to the Apple tree, she falleth into the pit and is taken.

" The herbe *Wolfebaine* or *Liberdine* is poison to Foxes, Wolves, Dogs, and Beares, and to all beasts that are littered blind, as the *Alpine Rhætians* affirme. There is one kinde of this called *Cyclamine*, which the *Valdensians* call *Tora*, and with the juice thereof they poison their darts, whereof I have credibly received this story; That a certain *Valdensian*, seeing a wilde beare, having a dart poysond heerewith, did cast it at the beare, being farre from him, and lightly wounded her, it being no sooner done, but the beare ran to and fro in a wonderful perplexitie through the woods, unto a verie sharpe cliffe of a rocke, where the man saw her draw her last breath, as soon as the poison entered to her hart, as he afterward found by opening of her bodie. The like is reported of henbane, another herb. But there is a certaine blacke fish in *Armenia* full of poison, with the pouder whereof they poison figs, and cast them in those places where wilde beastes are most plentifull, which they eat, and so are killed.

"Concerning the industrie or naturall disposition of a beare, it is certaine that they are very hardlie tamed, and not to be trusted though they seeme never so tame; for which cause there is a storie of *Diana* in *Lysias*, that there was a certaine beare made so tame, that it went uppe and downe among men, and woulde feede with them, taking meat at their handes, giving no occasion to feare or mistrust her cruelty; on a daye, a young mayde playing with the Beare, lasciviously did so provoke it, that he tore her in pieces; the Virgin's brethren seeing the murther, with their Dartes slew the Beare, whereupon followed a great pestilence through all that region: and when they consulted with the Oracle, the paynim God gave answeare, that the plague could not cease untill they dedicated some virginnes unto *Diana* for the Beare's sake that was slaine; which, some interpreting that they should sacrifice them, *Embarus*, upon condition the priesthoode might remaine in his family, slewe his onely daughter to end the pestilence, and for this cause the virgins were after dedicated to *Diana* before their marriage, when they were betwixt ten and fifteene yeare olde, which was performed in the moneth of *January*, otherwise they could not be married: yet beares are tamed for labours, and especially for sports among the *Roxalani* and *Libians*, being taught to draw water with wheeles out of the deepest wels; likewise stones upon sleds, to the building of wals.

"A prince of *Lituania* nourished a Beare very tenderly,

feeding her from his table with his owne hand, for he had used her to be familiar in his court, and to come into his owne chamber, when he listed, so that she would goe abroad into the fields and woods, returning home againe of her owne accord, and with her hand or foote rub the Kinge's chamber doore to have it opened, when she was hungry, it being locked. It happened that certaine young Noble men conspired the death of this Prince, and came to his chamber doore, rubbing it after the custome of the beare, the King not doubting any evill, and supposing it had bene his beare, opened the doore, and they presently slewe him. . . .

"There are many naturall operations in Beares. *Pliny* reporteth, that, if a woman bee in sore travaile of child-birth, let a stone, or arrow, which hath killed a man, a beare, or a bore, be throwne over the house wherein the Woman is, and she shall be eased of her paine. There is a small worme called *Volvox*, which eateth the vine branches when they are young, but if the vine-sickles be annointed with Beare's blood, that worme will never hurt them. If the blood or greace of a Beare be set under a bed, it will draw unto it all the fleas, and so kill them by cleaving thereunto. But the vertues medicinall are very many; and first of all, the blood cureth all manner of bunches and apostems in the flesh, and bringeth haire upon the eyelids if the bare place be annointed therewith.

"The fat of a Lyon is most hot and dry, and next to a Lyon's a Leopard's; next to a Leopard's a Beare's; and next to a Beare's, a Bul's. The later Physitians use it to cure convulsed and distracted parts, spots, and tumors in the body. It also helpeth the paine of the loins, if the sicke part be annointed therewith, and all ulcers in the legges or shinnes, when a plaister is made thereof with bole armoricke. Also the ulcers of the feet, mingled with allome. It is soveraigne against the falling of the haire, compounded with wilde roses. The Spaniards burne the braines of beares, when they die in any publicke sports, holding them venemous; because, being drunke, they drive a man to be as mad as a beare; and the like is reported of the heart of a Lyon, and the braine of a Cat. The right eie of a beare dried to pouder, and hung about children's neckes in a little bag, driveth away the terrour of dreames, and both the eyes whole, bound to a man's left arme, easeth a quartan ague.

"The liver of a sow, a lamb, and a bear put togither, and trod to pouder under one's shoos, easeth and defendeth cripples from inflamation: the gall being preserved and warmed in water, delivereth the bodie from Colde, when all other medicine faileth. Some give it, mixt with Water, to them that are bitten with a mad Dogge, holding it for a singular remedie, if the party can fast three daies before. It is also given against the palsie, the king's evill, the falling sickenesse, an old cough, the inflamation of the eies, the running of the eares, delevery in child birth, the Hæmorrhods, the weaknes of the backe, and the palsie: and that women may go their full time, they make ammulets of Bear's nails, and cause them to weare them all the time they are with Child."

The Fox.

By Englishmen, the Fox has been raised to the height of at least a demigod—and his cult is a serious matter attended with great minutiæ of ritual. Englishmen and Foxes cannot live together, but they live for one another, the man to hunt the fox, the fox to be hunted.

If there be a fox anywhere, even in the Campagna at Rome, and there are sufficient Englishmen to get up a scratch pack of hounds, there must "bold Reynard" be tortured with fear and exertion, only, in all probability, to die a cruel death in the end. In the Peninsular War, a pack of foxhounds accompanied the army; in India, failing foxes, they take the nearest substitute, the jackal; and in Australia, *faute de mieux*, they hunt the Dingo, or native dog. No properly constituted Englishman could ever compass the death of a poor fox, otherwise than by hunting. The Vulpecide—in any other manner—is, in an English county, a social leper—he is a thing *anathema*. Running away with a neighbour's wife may be condoned by county society, at least, among the men, but with them the man that shoots foxes is a very pariah, and it were good for that man had he never been born.

Every other nation, even from historic antiquity, has reckoned the Fox as among the ordinary *feræ naturæ*, to be killed, when met with, for the sake only of his skin, for his flesh is not toothsome: and when he arrives at the dignity of a silver or a black fox, his fur enwraps royal personages, as being of extreme value.

The Fox is noted everywhere for its "*craftiness*," and was so famed long before the epic of Reineke Fuchs was evolved, and, indeed, this may be said to be its principal attribute. Many are the stories told by country firesides of his stratagems, both in plundering and in his endeavours to escape from his enemies. Indeed, no country ought to be able to compare in Fox lore with our own. Its sagacity, cunning, or call it what you like, dates far back. Pliny tells us that "in Thrace, when all parts are covered with ice, the foxes are consulted, an animal, which, in other respects, is baneful from its

Craftiness. It has been observed, that this animal applies its ear to the ice, for the purpose of testing its thickness; hence it is, that the inhabitants will never cross frozen rivers and lakes, until the foxes have passed over them and returned."

The Fox is most abundant in the northern parts of Europe, and therefore we hear more about him from the pages of Olaus Magnus, Gessner, and Topsell.

The former says :—"When the fox is pressed with

hunger, Cold and Snow, and he comes near men's houses, he will bark like a dog, that house creatures may come nearer to him with more confidence. Also, he will faign himself dead, and lie on his back, drawing in his breath, and lolling out his tongue. The birds coming down, unawares, to feed on the carkasse, are snapt up by him, with open mouth. Moreover, when he is hungry, and finds nothing to eat, he rolls himself in red earth, that he may appear bloody; and, casting himself on the earth, he holds his breath, and when the birds see that he breaths not, and that his tongue hangs forth of his mouth, they think he is dead; but so soon as they descend, he draws them to him and devours them.

"Again, when he sees that he cannot conquer the Urchin, for his prickles, he lays him on his back, and so rends the soft part of his body. Sometimes fearing the multitude of wasps, he counterfeits and hides himself, his tail hanging out: and when he sees that they are all busie, and entangled in his thick tail, he comes forth, and rubs them against a stone or Tree, and kills them and eats them. The same trick, almost, he useth, when he lyes in wait for crabs and small fish, running about the bank, and he lets down his tail into the water, they admire at it, and run to it, and are taken in his fur, and pull'd out. Moreover, when he hath fleas, he makes a little bundle of soft hay wrapt in hair, and holds it in his mouth; then he goes by degrees into the water, beginning with his tail, that the fleas fearing the water, will run up all his body till they come at his head: then he dips in his head, that they may leap into the hay; when this is done, he leaves the hay in the water, and swims forth.

"But when he is hungry, he will counterfeit to play with the Hare, which he presently catcheth and devoureth, unlesse the Hare escape by flight, as he often doth. Sometimes he also escapes from the dogs by barking, faigning himself to be a dog, but more surely when he hangs by a bough, and makes the dogs hunt in vain to find his footing. He is also wont to deceive the Hunter and his dogs, when he runs among a herd of Goats, and goes for one of them, leaping upon the Goat's back, that he may sooner escape by the running of the Goat, by reason of the hatefull Rider on his back. The other Goats follow, which the Hunter fearing to molest, calls off his Dogs that many be not killed.

"If he be taken in a string, he will sometime bite off his own foot, and so get away. But, if there be no way open he will faign himself dead, that being taken out of the snare, he may run away. Moreover, when a dog runs after him, and overtakes him, and would bite him, he draws his bristly tail through the dog's mouth, and so he deludes the dog till he can get into the lurking places of the Woods. I saw also in the Rocks of *Norway* a Fox with a huge tail, who brought many Crabs out of the water, and then he ate them. And that is no rare sight, when as no fish like Crabs will stick to a bristly thing let down into the water, and to dry fish, laid on the rocks to dry. They that are troubled with the Gowt, are cured by laying the warm skin of this beast about the part, and binding it on. The fat, also, of the same creature, laid smeered upon the ears or lims of a gowty person, heals him; his fat is good for all torments of the guts, and for all pains, his brain often given to a child will preserve it ever from the Falling-sicknesse. These and such-like simple medicaments the North Country people observe."

A portion of the above receives a curious corroboration from Mr. P. Robinson in his book, *The Poets' Beasts*. Speaking of the Lynx, he says:—"But it is not, as is supposed, 'untamable.' The Gækwar of Baroda has a regular pack of trained lynxes, for stalking and hunting pea-fowl, and other kinds of birds. I have, myself, seen a tame lynx that had been taught to catch crows—no simple feat—and its strategy was as diverting as its agility amazing. It would lie down with the end of a string in its mouth, the other end being fast to a stake, and pretend to be asleep, dead asleep, drunk, chloroformed, anything you like that means profound and gross slumber. A foot or so off would be lying a piece of meat, or a bone.

"The crows would very soon discover the bone, and collecting round in a circle, would discuss the probabilities of the lynx only shamming, and the chances of stealing his dinner. The animal would take no notice whatever, but lie there looking so limp and dead, that at last one crow would make so bold as to come forward. The others let it do so alone, knowing that afterwards there would be a free fight for the plunder, and the thief, probably, not enjoy it, after all. So the delegate would advance with all the caution of a crow—and nothing exceeds it—until within seizing distance. There it would stop, flirt its wings nervously, stoop, take a last long look at the lynx to make sure that it really *was* asleep, and then dart like lightning at the bone. But, if the crow was as quick as lightning, the lynx was as swift as thought, and lo! the next instant there was the beast sitting up with the bird in its mouth! . . .

"Next time it had to practise a completely different manœuvre. The same crows are not to be 'humbugged' a second time by a repetition of the being-dead trick. So the lynx, when a sufficient number of the birds had assembled, would take the string in its mouth, and run round and round the stake, at the extreme limit of its tether, as if it were tied. The crows, after their impudent fashion, would close in. They thought they knew the exact circumference of the animal's circle, and getting as close to the dangerous line as possible, without actually transgressing it, would mock and abuse the supposed be-tethered brute. But all of a sudden, the circling lynx would fly out at a tangent, right into the thick of his black tormentors, and, as a rule, bag a brace, right and left."

Topsell gives some curious particulars of the Fox, and, speaking of their earths, he says:—"These dens have many caves in them, and passages in and out, that when the Terrars shall set upon him in the earth, he may go forth some other way, and forasmuch as the Wolfe is an enemy to the Foxe, he layeth in the mouth of his den, an Herbe (called Sea-onyon) which is so contrary to the nature of a Wolfe, and he so greatly terrified therewith, that hee will never come neere the place where it groweth, or lyeth; the same is affirmed of the Turtle to save her young ones, but I have not read that Wolves will prey upon Turtles, and therefore we reject that as a fable. . . . If a Foxe eat any meat wherein are bitter Almondes, they die thereof, if they drinke not presently: and the same thing do Aloes in their meate worke uppon them, as *Scaliger* affirmeth upon his owne sighte or knowledge. *Apocynon* or Bear-foot given to dogs, wolves, Foxes, and all other beasts which are littered blind, in fat, or any other meat, killeth them, if vomit helpe them not, which falleth out very seldome, and the seeds of this hearbe have the same operation. It is reported by *Democritus*, that, if wilde rue be secretly hunge under a Hen's wing, no Fox will meddle with her, and the same writer also declareth for approoved, that, if you mingle the gal of a Fox, or a Cat, with their ordinary foode, they shall remaine free from the danger of these beasts.

"The medicinall uses of this beast are these: first, (as *Pliny*, and *Marcellus* affirme) a Fox sod in water until nothing of the Foxe be left whole except the bones, and the Legges, or other parts of a gouty body, washed, and daily bathed therein, it shall drive away all paine and griefe strengthening the defective and weake members; so also it cureth all the shrinking up and paines in the sinnewes: and *Galen* attributeth the same vertue to an *Hyæna* sod in Oyle, and the lame person bathed therein, for it hath such power to evacuate and draw forth whatsoever evill humour aboundeth in the body of man, that it leaveth nothing hurtfull behinde.

"Neverthelesse, such bodies are soon againe replenished through evill dyet, and relapsed into the same disease againe. The Fox may be boyled in fresh or salt water with annise and time, and with his skin on whole, and not slit, or else his head cut off, there being added to the decoction two pintes of oyle.

"The flesh of a Foxe sod and layed to afore bitten by a Sea hare, it cureth and healeth the same. The Foxe's skinne is profitable against all moyste fluxes in the skinne of the body, and also the gowt, and cold in the sinnewes. The ashes of Foxe's flesh burnt and drunk in wine, is profitable against the shortnesse of breath and stoppings of the liver.

"The blood of a Foxe dissected, and taken forth of his urine alive, and so drunk, breaketh the stone in the bladder, or else (as *Myrepsus* saieth) kill the Foxe, and take the blood, and drink a Cupfull thereof, and afterward with the same wash the parts, and, within an houre the stone shall be voyded: the same vertue is in it being dryed and drunke in wine with sugar.

"*Oxycraton* and Foxes blood infused into the Nostrils of a lethargick Horsse, cureth him. The fat is next to a Bul's and a Swine's, so that the fat or larde of Swine may be used for the fat of Foxes, and the fat of Foxes for the Swines grease in medicine. Some do herewith annoynte the places which have the Crampe, and all trembling and shaking members. The fatte of a Foxe and a Drake enclosed in the belly of a Goose, and so rosted, with the dripping that commeth from it, they annoynt paralyticke members.

"The same, with powder of Vine twigs mollified and sod in lye, attenuateth, and bringeth downe, all swelling tumours of the flesh. The fat alone healeth the *Alopecias* and looseness of the haire; it is commended in the cure of all sores and ulcers of the head, but the gall, and time, with Mustard-seede is more approved. The fat is also respected for the cure of paine in the eares, if it be warmed and melt at the fire, and so instilled; and this is used against tingling in the eares. If the Haires rot away on a Horse's taile, they recover them againe, by washing the place with urine and branne, with Wyne and Oyle, and afterward annoynt it with foxe's grease. When sores or ulcers have procured the haire to fall off from the heade, take the head of a young foxe burned with the leaves of blacke *Orchanes* and *Alcyonium*, and the powder cast upon the head recovereth againe the haire.

"If the braine be often given to infants and sucking children, it maketh them that they shall remaine free from the falling evill. *Pliny* prescribeth a man which twinkleth with his eies, and cannot looke stedfastly, to weare in a chaine, the tongue of a foxe; and *Marcellus* biddeth to cut out the tongue of a live foxe, and to turne him away, and hang uppe that tongue to dry in purple thred, and, afterward put it about his necke that is troubled with the whitenesse of the eies, and it shall cure him.

"But it is more certainely affirmed, that the tongue, either dryed, or greene, layed to the flesh wherein is any Dart or other sharpe head, it draweth them forth violently, and rendeth not the flesh, but, only where it is entred. The liver dryed, and drunke cureth often sighing. The same, or the lights drunke in blacke Wine, openeth the passages of breathing. The same washed in Wyne, and dryed in an earthen pot in an Oven, and, afterward, seasoned with Sugar, is the best medicine in the world for an old cough, for it hath bin approved to cure it, although it hath continued twenty years, drinking every day two sponfuls in Wine.

"The lightes of foxes drunke in Water after they have beene dryed into powder, helpeth the Melt, and *Myrepsus* affirmeth, that when he gave the same powder to one almost suffocated in a pleurisie it prevailed for a remedy. *Archigene* prescribeth the dried liver of a Fox for the Spleneticke with Oxymell: and *Marcellinus* for the Melt, drunke after the same manner; and *Sextus* adviseth to drinke it simply without composition of Oxymell. The gall of a Foxe instilled into the eares with Oyle, cureth the paine in them, and, mixed with Hony Atticke, and annointed upon the eies, taketh away al dimnes from them, after an admirable manner. The melt, bound upon the tumors, and bunches of the brest, cureth the Melt in man's body. The reynes dried and mingled with Honie, being anointed uppon Kernels, take them away. For the swelling of the Chaps, rub the reines of a Fox within the mouth. The dung, pounded with Vineger, by annointment cureth the Leprosie speedily. These and such other vertues medicinal, both the elder and later Phisitians have observed in a Fox,—wherewithal we wil conclude this discourse."

THE WOLF.

The Wolf, as a beast of prey, is invested with a terror peculiarly its own; when solitary, it is not much dreaded by, and generally shrinks from, man, but, united by hunger into packs, they are truly to be dreaded, for they spare nor man nor beast. They lie, too, under the imputation of magic, and have done so from a very early age. Their cunning, instinct, or reasoning powers, are almost as well developed as in the fox, and, of all the authorities I have consulted, the one best fitted to discourse upon the Wolf and his peculiarities is Topsell, and here is one of their idiosyncrasies :—

"It is said that Wolves doe also eate a kind of earth called *Argilla*, which they doe not for hunger, but to make their bellies waigh heavy, to the intent, that when they set upon a Horsse, an Oxe, a Hart, an Elke, or some such strong beast, they may waigh the heavier, and hang fast at their throates till they have pulled them downe, for by vertue of that tenacious earth, their teeth are sharpened, and the waight of their bodies encreased; but, when they have killed the beast that they set upon, before they touch any part of his flesh, by a kind of natural vomit, they disgorge themselves, and empty their bellies of the earth, as unprofitable food. . . .

"They also devoure Goates and Swyne of all sortes, except Bores, who doe not easily yeald unto Wolves. It is said that a Sow, hath resisted a Wolfe, and when he fighteth with her, hee is forced to use his greatest craft and suttelty, leaping to and from her with his best activity, least she should lay her teeth upon him, and so at one time deceive him of his prey, and deprive him of his life. It is reported of one that saw a Wolfe in a Wood, take in his mouth a peece of Timber of some thirty or forty pound waight, and with that he did practise to leape over the trunke of a tree that lay upon the earth; at length, when he perceived his own ability and dexterity in leaping with that waight in his mouth, he did there make his cave, and lodged behinde that tree; at last, it fortuned there came a wild Sow to seeke for meat along by that tree, with divers of her pigs following her, of different age, some a yeare olde, some halfe a yeare, and some lesse. When he saw them neare him, he suddenly set upon one of them, which he conjectured was about the waite of Wood which he carried in his mouth, and when he had taken him, whilest the old Sow came to deliver her pig at his first crying, he suddenly leaped over the tree with the pig in his mouth, and so was the poore Sow beguiled of her young one, for she could not leape after him, and yet might stand and see the Wolfe to eate the pigge, which hee had taken from her. It is also sayd, that when they will deceive Goates, they come unto them with the greene leaves and small boughes of Osiers in their mouthes, wherewithall they know Goats are delighted, that so they may draw them therewith, as to a baite, to devour them.

"Their maner is, when they fal upon a Goat or a Hog, or some such other beast of smal stature, not to kil them, but to lead them by the eare with al the speed they can drive them, to their fellow Wolves, and, if the beast be stubborne, and wil not runne with him, then he beateth his hinder parts with his taile, in the mean time holding his ear fast in his mouth, whereby he causeth the poore beast to run as fast, or faster than himselfe unto the place of his owne execution, where he findeth a crew of ravening Wolves to entertaine him, who, at his first appearance seize upon him, and, like Divels teare him in peeces in a moment, leaving nothing uneaten but onely his bowels. . . .

"Now although there be a great difference betwixt him and a Bul, both in strength and stature, yet he is not affraid to adventure combat, trusting in his policy more than his vigor, for when he setteth upon a Bul, he commeth not upon the front for feare of his hornes, nor yet behind him for feare of his heeles, but first of al standeth a loofe from him, with his glaring eyes, daring and provoking the Bul, making often profers to come neere unto him, yet is wise enough to keepe a loofe till he spy his advauntage, and then he leapeth suddenly upon the backe of the Bul at the one side, and being so ascended, taketh such hold, that he killeth the beast, before he loosen his teeth. It is also worth the observation, how he draweth unto him a Calfe that wandereth from the dam, for by singular treacherie he taketh him by the nose, first drawing him forwarde, and then the poore beast striveth and draweth backward, and thus they struggle togither, one pulling one way, and the other another, till at last the Wolfe perceiving advantage, and feeling when the Calfe pulleth heavyest, suddenly he letteth go his hold, whereby the poore beast falleth backe upon his buttocks, and so downe right upon his backe; then flyeth the Wolfe to his belly which is then his upper part, and easily teareth out his bowels, so satisfieng his hunger and greedy appetite.

"But, if they chance to see a Beast in the water, or in the marsh, encombred with mire, they come round about him, stopping up al the passages where he shold come out, baying at him, and threatning him, so as the poore distressed Oxe plungeth himselfe many times over head and eares, or at the least wise they so vex him in the mire, that they never suffer him to come out alive. At last, when they perceive him to be dead, and cleane without life by suffocation, it is notable to observe their singular subtilty to drawe him out of the mire, whereby they may eat him ; for one of them goeth in, and taketh the beast by the taile, who draweth with al the power he can, for wit without strength may better kill a live Beast, than remove a dead one out of the mire; therefore, he looketh behind him, and calleth for more helpe ; then, presently another of the wolves taketh that first wolve's tail in his mouth, and a third wolf the second's, a fourth the third's, a fift the fourth, and so forward, encreasing theyr strength, until they have pulled the beast out into the dry lande. *Sextus* saith that, in case a Wolf do see a man first, if he have about him the tip of a Wolf's taile, he shal not neede to feare anie harme. All domestical Foure footed beasts, which see the eie of a wolfe in the hand of a man, will presently feare and runne away.

" If the taile of a wolfe be hung in the cratch of Oxen, they can never eat their meate. If a horse tread upon the foote steps of a Wolfe, which is under a Horse-man or Rider, hee breaketh in peeces, or else standeth amazed. If a wolfe treadeth in the footsteps of a horse which draweth a waggon, he cleaveth fast in the rode, as if he were frozen.

" If a Mare with foale, tread upon the footsteps of a wolfe, she casteth her foal, and therefore the Egyptians, when they signifie abortment doe picture a mare treading upon a wolf's foot. These and such other things are reported, (but I cannot tell how true) as supernaturall accidents in wolves. The wolfe also laboureth to overcome the Leoparde, and followeth him from place to place, but, for as much as they dare not adventure upon him single, or hand to hand, they gather multitudes, and so devoure them. When wolves set upon wilde Bores, although they bee at variance amonge themselves, yet they give over their mutual combats, and joyne together against the Wolfe their common adversarie.

"And this is the nature of this beast, that he feareth no kind of weapon except a stone, for, if a stone be cast at him, he presently falleth downe to avoide the stroke, for it is saide that in that place of his body where he is wounded by a stone, there are bred certaine wormes which doe kill and destroie him. . . . As the Lyon is afraide of a white Cocke and a Mouse, so is the wolfe of a Sea crab, or shrimp. It is said that the pipe of *Pithocaris* did represse the violence of wolves when they set upon him, for he sounded the same unperfectly, and indistinctly, at the noise whereof the raging wolfe ran away; and it hath bin beleeved that the voice of a singing man or woman worketh the same effect.

"Concerning the enimies of wolves, there is no doubt but that such a ravening beast hath fewe friends, . . . for this cause, in some of the inferiour beasts their hatred lasteth after death, as many Authors have observed; for, if a sheepe skinne be hanged up with a wolves's skin, the wool falleth off from it, and, if an instrument be stringed with stringes made of both these beasts the one will give no sounde in the presence of the other."

Here we have had all the bad qualities of the Wolf depicted in glowing colours; but, as a faithful historian, I must show him also under his most favourable aspect—notably in two instances—one the she-wolf that suckled Romulus and Remus, and the other who watched so tenderly over the head of the Saxon Edmund, King and Martyr, after it had been severed from his body by the Danes, and contemptuously thrown by them into a thicket.

His mourning followers found the body, but searched for some time for the head, without success; although they made the woods resound with their cries of "Where artow, Edward?" After a few days' search, a voice answered their inquiries, with "Here, here, here." And, guided by the supernatural voice, they came upon the King's head, surrounded by a glory, and watched over, so as to protect it from all harm—by a *WOLF!* The head was applied deftly to the body, which it joined naturally; indeed, so good a job was it, that the junction could only be perceived by a thin red, or purple, line.

It must be said of this wolf, that he was *thorough*, for not content with having preserved the head of the Saintly King from harm, he meekly followed the body to St. Edmund's Bury, and waited there until the funeral; when he quietly trotted back, none hindering him, to the forest.

Were-Wolves.

But of all extraordinary stories connected with the Wolf, is the belief which existed for many centuries, (and in some parts of France still does exist, under the form of the "Loup-garou,") and which is mentioned by many classical authors—Marcellus Sidetes, Virgil, Herodotus, Pomponius Mela, Ovid, Pliny, Petronius, &c.—of men being able to change themselves into wolves. This was called *Lycanthropy*, from two Greeks words signifying wolf, and man, and those who were thus gifted, were dignified by the name of *Versipellis*, or able to change the skin. It must be said, however, for Pliny, amongst classical authors, that although he panders sufficiently to popular superstition to mention Lycanthropy, and quotes from others some instances of it, yet he writes:—

"It is really wonderful to what a length the credulity of the Greeks will go! There is no falsehood, if ever so barefaced, to which some of them cannot be found to bear testimony."

This curious belief is to be found in Eastern writings, and it was especially at home with the Scandinavian and Teutonic nations. It is frequently mentioned in the Northern Sagas—but space here forbids more than just saying that the best account of these *eigi einhamir* (not of one skin) is to be found in *The Book of Were-Wolves*, by the Rev. S. Baring-Gould.

The name of *Were Wolf*, or *Wehr Wolf*, is derived thus, according to Mr. Gould:—" *Vargr* is the same as *u-argr*, restless ; *argr* being the same as the Anglo-Saxon *earg*. *Vargr* had its double signification in Norse. It signified a Wolf, and also a godless man. This *vargr* is the English *were*, in the word were-wolf, and the *garou* or *varou* in French. The Danish word for were-wolf is *var-ulf*, the Gothic, *vaira-ulf*." Lycanthropy was a widespread belief, but it gradually dwindled down in the sixteenth and seventeenth centuries to those *eigi einhamir*, the witches who would change themselves into hares, &c.

Olaus Magnus tells us *Of the Fiercenesse of Men who by Charms are turned into Wolves*:—" In the Feast of Christ's Nativity, in the night, at a certain place, that they are resolved upon amongst themselves, there is gathered together such a huge multitude of Wolves changed from men, that dwell in divers places, which afterwards the same night doth so rage with wonderfull fiercenesse, both against mankind, and other creatures that are not fierce by nature, that the Inhabitants of that country suffer more hurt from them than ever they do from the true natural Wolves. For as it is proved, they set upon the houses of men that are in the Woods, with wonderfull fiercenesse, and labour to break down the doors, whereby they may destroy both men and other creatures that remain there.

"They go into the Beer-Cellars, and there they drink out some Tuns of Beer or Mede, and they heap al the empty vessels one upon another in the midst of the Cellar, and so leave them: wherein they differ from natural and true Wolves. But the place, where, by chance they stayd that night, the Inhabitants of those Countries think to be prophetical: Because, if any ill successe befall a Man in that place; as, if his Cart overturn, and he be thrown down in the Snow, they are fully perswaded that man must die that year, as they have for many years proved it by experience. Between *Lituania*, *Samogetia*, and *Curonia*, there is a certain wall left, of a Castle that was thrown down; to this, at a set time, some thousands of them come together, that each of them may try his nimblenesse in leaping. He that cannot leap over this wall, as commonly the fat ones cannot, are beaten with whips by their Captains.

"And it is constantly affirmed that amongst that multitude there are the great men, and chiefest Nobility of the Land. The reason of this metamorphosis, that is exceeding contrary to Nature, is given by one skilled in this witchcraft, by drinking to one in a Cup of Ale, and by mumbling certain words at the same time, so that he who is to be admitted into that unlawful Society, do accept it. Then, when he pleaseth, he may change his humane form, into the form of a Wolf entirely, going into some private Cellar, or secret Wood. Again, he can, after some time put off the same shape he took upon him, and resume the form he had before at his pleasure. . . .

"But for to come to examples; When a certain Nobleman took a long journey through the Woods, and had many servile Country-fellows in his Company, that were acquainted with this witchcraft, (as there are many such found in those parts) the day was almost spent; wherefore he must lie in the Woods, for there was no Inne neare that place; and withall they were sore pinched with hunger and want. Last of all, one of the Company propounded a seasonable proposall, that the rest must be quiet, and if they saw any thing they must make no tumulte; that he saw afar off a flock of sheep feeding; he would take care that, without much labor, they should have one of them to rost for Supper. Presently he goes into a thick Wood that no man might see him, and there he changed his humane shape like to that of a Wolf. After this he fell upon the flock of sheep with all his might, and he took one of them that was running back to the Wood, and then he came to the Chariot in the form of a Wolf, and brought the sheep to them. His companions being conscious how he stole it, receive it with grateful mind, and hide it close in the Chariot; but he that had changed himself into a Wolf, went into the Wood again, and became a Man.

"Also in *Livonia* not many years since, it fell out that there was a dispute between a Nobleman's wife and his servant, (of which they have plenty more in that Country, than in any Christian Land) that men could not be turned into Wolves; whereupon he brake forth into this speech, that he would presently shew her an example of that businesse, so he might do it with her permission: he goes alone into the cellar, and, presently after, he came forth in the form of a Wolf. The dogs ran after him through the fields to the wood, and they bit out one of his eyes, though he defended himself stoutly enough. The next day he came with one eye to his Lady. Lastly, as is yet fresh in memory, how the Duke of *Prussia*, giving small credit to such a Witchcraft, compelled one who was cunning in this Sorcery, whom he held in chains, to change himself into a Wolf; and he did so. Yet that he might not go unpunished for this Idolatry, he afterwards caused him to be burnt. For such heinous offences are severely punished both by Divine and Humane Laws."

Zahn, on the authority of Trithemius, who wrote in 1335, says that men having the spine elongated after the manner of a tail were Were-wolves. Topsell takes a more sensible view of the matter:—" There is a certaine territorie in Ireland (whereof M. *Cambden* writeth) that the inhabitants which live till they be past fifty yeare old, are foolishly reported to be turned into wolves, the true cause whereof he conjectureth to be, because for the most part they are vexed with the disease called *Lycanthropia*, which is a kind of melancholy, causing the persons so affected, about the moneth of February, to forsake their owne dwelling or houses, and to run out into the woodes, or neare the graves and sepulchers of men, howling and barking like Dogs and Wolves. The true signes of this disease are thus described by *Marcellus*: those, saith he, which are thus affected, have their faces pale, their eies dry and hollow, looking drousily and cannot weep. Their tongue as if it were al scab'd, being very rough, neither can they spit, and they are very thirsty, having many ulcers breaking out of their bodies, especially on their legges; this disease some cal

Lycaon, and men oppressed therewith, *Lycaones*, because that there was one *Lycaon*, as it is fained by the poets, who, for his wickednes in sacrificing of a child, was by *Jupiter* turned into a Wolf, being utterly distracted of human understanding, and that which the poets speake of him. And this is most strange, that many thus diseased should desire the graves of the dead."

The Antelope.

When not taken from living specimens, or skins, the artists of old drew somewhat upon their imaginations for

their facts, as is the case with this Antelope, of which Topsell gives the following description:—"They are bred in *India*, and *Syria*, neere the River *Euphrates*, and delight much to drinke of the cold water thereof. Their bodie is like the body of a *Roe*, and they have hornes growing forthe of the crowne of their head, which are very long and sharpe; so that *Alexander* affirmed that they pierced through the sheeldes of his Souldiers, and fought with them very irefully: at which time his company slew as he travelled to *India*, eight thousand, five hundred, and fifty; which great slaughter may be the occasion why they are so rare, and seldome seene to this day, by cause thereby the breeders, and meanes of their continuance (which consisted in their multitude) were weakened and destroyed. Their hornes are great, and made like a saw, and they, with them, can cut asunder the braunches of *Osier*, or small trees, whereby it commeth to passe that many times their necks are taken in the twists of the falling boughes, whereat the Beast with repining cry, bewrayeth himselfe to the Hunters, and so is taken. The vertues of this Beast are unknowne, and therefore *Suidas* sayth an *Antalope* is but good in part."

The Horse.

Aldrovandus gives us a curious specimen of a horse, which the artist has drawn with the slashed trunk breeches of the time. He says that *Fincelius*, quoting *Licosthenes*, mentions that this animal had its skin thus slashed, from its birth, and was to be seen about the year 1555. Its skin was as thick as sole-leather. It was, probably, an ideal Zebra.

Topsell gives us some fine horse-lore, especially as to their love for their masters:—" *Homer* seemeth also to affirme that there are in Horsses divine qualityes, understanding things to come, for, being tyed to their mangers they mournd for the death of *Patroclus*, and also shewed *Achilles* what should happen unto him; for which cause *Pliny* saieth of them that they lament their lost maisters with teares, and foreknow battailes. *Accursius* affirmeth that *Cæsar* three daies before he died, found his ambling Nag weeping in the stable,

which was a token of his ensewing death, which thing I should not beleeve, except *Tranquillus* in the life of *Cæsar*, had related the same thing, and he addeth moreover, that the Horsses which were consecrated to *Mars* for passing over *Rubicon*, being let to run wilde abroad, without their maisters, because no man might meddle with the horses of the Gods, were found to weepe abundantly, and to abstaine from all meat.

"Horsses are afraid of Elephants in battaile, and likewise of a Cammell, for which cause when *Cyrus* fought against *Cræsus*, he overthrew his Horse by the sight of Camels, for a horse cannot abide to looke upon a Camell. If a Horse tread in the footpath of a Wolfe, he presently falleth to be astonished; Likewise, if two or more drawing a Charriot, come into the place where a Wolfe hath trod, they stand so still as if the Charriot and they were frozen to the earth, sayth *Ælianus* and *Pliny*. *Æsculapius* also affirmeth the same thing of a Horsse treading in a Beare's footsteppes, and assigneth the reason to be in some secret, betweene the feete of both beastes. . . .

"Al kind of Swine are enemies to Horses, the Estridge also, is so feared of a Horse, that the Horsse dares not appeare in his presence. The like difference also is betwixt a Horse, and a Beare. There is a bird which is called *Anclorus*, which neyeth like a Horse, flying about; the Horse doth many times drive it away; but because it is somewhat blind, and cannot see perfectly, therefore the horsse doth oftentimes ketch it, and devoure it, hating his owne voice in a creature so unlike himself.

"It is reported by *Aristotle*, that the Bustard loveth a Horsse exceedingly, for, seeing other Beastes feeding in the pastures, dispiseth and abhorreth them; but, as soone as ever it seeth a Horsse, it flyeth unto him for joy, although the Horsse run away from it: and, therefore, the Egyptians, when they see a weake man driving away a stronger, they picture a Bustard flying to a Horsse. . . .

"*Julius Cæsar* had a horsse which had cloven hooves like a man's fingers, and because he was foaled at that time when the sooth-sayers had pronounced that hee should have the government of the world, therefore he nourished him carefully, and never permitted any man to backe him but himselfe, which he afterwards dedicated in the Temple of Venus. . . .

"If one do cut the vaines of the pallet of a horse's mouth, and let it runne downe into his belly, it will presently destroy and consume the maw, or belly worms, which are within him. The Marrow of a horse is also very good to loosen the sinewes which are knit and fastned together, but first let it be boiled in wine, and afterwards be made cold, and then anointed warmly either by the Fire, or Sun. The teeth of a male horse not gelded, or by any labor made feeble, being put under the head, or over the head of him that is troubled or startleth in his dreame, doth withstand and resist all unquietnes which in the time of his rest might happen unto him. The teeth also of a horse is verye profitable for the curing of the Chilblanes which are rotten and full of corruption when they are swollen full ripe. The teeth which do, first of all, fall from horses, being bound or fastned upon children in their infancie, do very easily procure the breeding of the teeth, but with more speed, and more effectually, if they have never touched the ground. . . .

"If you anoint a combe with the foame of a horse, wherewith a young man or youth doth use to comb his head, it is of such force as it will cause the haire of his head neither to encrease or any whit to appeare. The foame of a horse is also very much commended for them which have either pain or difficulty of hearing in their ears, or else the dust of horse dung, being new made and dryed, and mingled with oyle of Roses. The griefe or soreness of a man's mouth or throat, being washed or annointed with the foame of a Horse, which hath bin fed with Oates or barly, doth presently expell the paine of the Sorenesse, if so be that it be 2 or 3 times washed over with the juyce of young or greene Sea-crabs beaten small together." But I could fill pages with remedial recipes furnished by the horse.

The Mimick Dog.

"The Mimicke or Getulian Dogge," is, I take it, meant for a poodle. It was "apt to imitate al things it seeth, for which cause some have thought that it was conceived by an Ape, for in wit and disposition it

resembleth an Ape, but in face, sharpe and blacke like an Hedgehog, having a short recurved body, very long legs, shaggy haire, and a short taile: this is called of some *Canis Lucernarius*. These being brought up with apes in their youth, learne very admirable and strange feats, whereof there were great plenty in *Egypt* in the time of king *Ptolemy*, which were taught to leap, play, and dance, at the hearing of musicke, and in many poore men's houses they served insteed of servaunts for divers uses.

"These are also used by Plaiers and Puppet-Mimicks to worke straunge trickes, for the sight whereof they get much money; such an one was the Mimick's dog, of which *Plutarch* writeth that he saw in a publicke spectacle at Rome before the Emperor *Vespasian*. The dog was taught to act a play, wherein were contained many persons' parts, I mean the affections of many other dogs; at last, there was given him a piece of bread, wherein, as was saide, was poison, having vertue to procure a dead sleepe, which he received and swallowed; and presently, after the eating thereof, he began to reele and stagger too and fro like a drunken man, and fell downe to the ground, as if he had bin dead, and so laie a good space, not stirring foot nor lim, being drawne uppe and downe by divers persons, according as the gesture of the play he acted did require, but when he perceived by the time, and other signes that it was requisite to arise, he first opened his eies, and lift up his head a little, then stretched forth himself, like as one doth when he riseth from sleepe; at last he geteth up, and runneth to him to whom that part belonged, not without the joy, and good content of *Cæsar* and all other beholders.

"To this may be added another story of a certaine Italian about the yeare 1403, called *Andrew*, who had a red Dog with him, of strange feats, and yet he was blind. For standing in the Market place compassed about with a circle of many people, there were brought by the standers by, many Rings, Jewels, bracelets, and peeces of gold and silver, and these, within the circle were covered with earth, then the dog was bid to seeke them out, who with his nose and feet did presently find and discover them, then was hee also commaunded to give to every one his owne Ring, Jewell, Bracelet, or money, which the blind dog did performe directly without stay or doubt. Afterward, the standers by, gave unto him divers pieces of coine, stamped with the images of sundry princes, and then one of them called for a piece of English money, and the Dog delivered him a piece; another for the Emperor's coine, and the dog delivered him a piece thereof; and so consequently, every princes coine by name, till all was restored; and this story is recorded by Abbas Urspergensis, where upon the common people said, the dog was a divell, or else possessed with some pythonicall spirit."

It is curious to note some of the remedies against hydrophobia—and I only give a portion of the long list.

"For the outward compound remedies, a plaister made of *Opponax* and Pitch, is much commended, which *Menippus* used, taking a pound of Pitch of Brutias, and foure ounces of *Opponax*, adding withall, that the *Opponax* must be dissolved in vinegar, and afterwards the Pitch and the vinegar must be boiled together, and when the vinegar is consumed, then put in the *Opponax*, and of both together make like taynters or splints, and thrust them into the wound, so let them remaine many dayes together, and in the meane time drinke an antidot of sea crabs and vineger, (for vineger is alway pretious in this confection). Other use *Basilica*, Onyons, Rue, Salt, Rust of Iron, white bread, seedes of hore hound, and

CURIOUS CREATURES.

triacle: but the other plaister is most forcible to be applyed outwardly, above al medicines in the world.

"For the simple or uncompounded medicines to be taken against this sore, are many: As Goose-grease, the roote of Wilde roses drunke; bitter Almonds, leaves of Chickweed, or Pimpernell, the old skinne of a snake pounded with a male sea Crab, Betony, Cabbage-leaves, or stalkes, with Persneps and vineger, lime and sewet, poulder of Sea-Crabs with Hony; poulder of the shels of Sea-Crabs, the haires of a Dog layed on the wound, the head of the Dog which did bite, mixed with a little *Euphorbium;* the haire of a man with vineger, dung of Goates with wine, Walnuts with Hony and salte, poulder of fig tree in a sear cloth, Fitches in wine, *Euphorbium*, warme horse-dung, raw beanes chewed in the mouth, fig tree leaves, greene figs with vineger, fennel stalkes, Gentians, dung of pullen, the Lyver of a Buck-goate, young swallowes, burned to poulder, also their dung; the urine of a man, an Hyæna's skin, flower de luce with honey, a Sea hearb called *Kakille*, *Silphum* with salt, the flesh and shels of snayles, leeke seeds with salt, mints, the taile of a field mouse cut off from her alive, and she suffered to live, rootes of Burres, with salt of the Sea plantaine, the tongue of a Ramme with salt, the flesh of al Sea-fishes, the fat of a sea-Calfe and Vervine, besides many other superstitious amulets which are used to be bound to the Armes, neckes, and brests, as the Canine tooth bound up in a leafe, and tyed to the Arme. A worme bred in the dung of Dogges, hanged about the necke, the roots of *Gentian* in an Hyæna's skin, or young Wolfe's Skin, and such like; whereof I know no reason beside the opinion of men."

Let us now see what medicinal properties exist in dogs

themselves; and, here again, I must very much curtail the recital of their benefits to mankind.

"The vertues of a Dog's head made into poulder, are both many and unspeakable, by it is the biting of mad dogs cured, it cureth spots, and bunches in the head, and a plaister thereof made with Oyle of Roses, healeth the running in the head. The poulder of the teeth of Dogges, maketh Children's teeth to come forth with speed and easie, and, if their gums be rub'd with a dog's tooth, it maketh them to have the sharper teeth; and the poulder of these Dogs teeth rubbed upon the Gummes of young or olde, easeth toothache, and abateth swelling in the gummes. The tongue of a Dogge, is most wholesome both for the curing of his owne wounds by licking, as also of any other creature. The rennet of a Puppy drunke with Wine, dissolveth the Collicke in the same houre wherein it was drunke," &c., &c., &c.

The Cat.

Aldrovandus gives us a picture of a curly-legged Cat, but, beyond saying that it was so afflicted (or ornamented) from its birth, he gives no particulars. Topsell, too, is singularly silent on the merits of Cats; but yet he mentions some interesting particulars respecting them:—" To keepe Cats from hunting of Hens, they use to tie a little wild rew under their wings, and so likewise from Dovecoates, if they set it in the windowes, they dare not approach unto it for some secret in nature. Some have said that cats will fight with Serpentes, and Toads, and kill them, and, perceiving that she is hurt by them, she presently drinketh water, and is cured: but I cannot consent unto this opinion. . . . *Ponzettus* sheweth by experience that cats and Serpents love one another, for there was (sayth he) in a certain Monastery, a Cat norished by the Monkes, and suddenly the most part of the Monkes which used to play with the Cat, fell sicke; whereof the Physitians could find no cause, but some secret poyson, and al of them were assured that they never tasted any: at the last a poore laboring man came unto them, affirming that he saw the Abbey-Cat playing with a Serpent, which the Physitians understand-

ing, presently conceived that the Serpent had emptied some of her poyson upon the Cat, which brought the same to the Monkes, and they by stroking and handeling the Cat, were infected therewith; and whereas there remained one difficulty, namely, how it came to passe the Cat herself was not poisoned thereby, it was resolved, that, forasmuch as the Serpentes poison came from him but in playe and sporte, and not in malice and wrath, that therefore the venom thereof being lost in play, neither harmed the Cat at al, nor much endangered the Monkes; and the very like is observed of Myce that will play with Serpents. . . .

"Those which will keepe their Cattes within doores, and from hunting Birds abroad, must cut off their eares, for they cannot endure to have drops of raine distil into them, and therefore keep themselves in harbor. . . . They cannot abide the savour of oyntments, but fall madde thereby; they are sometimes infected with the falling evill, but are cured with *Gobium*."

The Lion.

Of the great Cat, the Lion, the ancients give many wonderful stories, some of them not altogether redounding to his character for bravery:—"A serpent, or snake doth easily kill a lion, where of *Ambrosius* writeth very elegantly. *Eximia leonis pulchritudo, per comantes cervicis toros excutitur, cum subito a serpente os pectore tenus attolitur, itaque Coluber cervum fugit sed Leonem interficit. The splendant beautie of a lion in his long curled mane is quickly abated, and allayed, when the serpent doth but lift up his head to his brest.* For such is the ordinance of God, that the Snake, which runneth from a fearefull Hart, should without all feare kill a courageous Lyon; and the writer of Saint *Marcellus* life, *How much more will he feare a great Dragon, against whom he hath not power to lift up his taile.* And *Aristotle* writeth that the Lyon is afraid of the Swine, and *Rasis* affirmeth as much of the mouse.

"The Cocke also both seene and heard for his voice and combe, is a terror to the Lion and Basiliske, and the Lyon runneth from him when he seeth him, especially from a white cocke, and the reason hereof, is because they are both partakers of the Sunnes qualities in a high degree, and therefore the greater body feareth the lesser, because there is a more eminent and predominant sunny propertie in the Cocke, than in the Lion. *Lucretius* describes this terrour notably, affirming that, in the morning, when the Cocke croweth, the lions betake themselves to flight, because there are certain seedes in the body of Cockes, which when they are sent, and appear to the eyes of Lions, they vexe their pupils and apples, and make them, against Nature, become gentle and quiet."

The Leontophonus—The Pegasus—The Crocotta.

The Lion has a dreadful enemy, according to Pliny, who says:—"We have heard speak of a small animal to which the name of *Leontophonus*[1] has been given, and which is said to exist only in those countries where the Lion is produced. If its flesh is only tasted by the Lion, so intensely venomous is its nature, that this lord of the other quadrupeds instantly expires. Hence it is that the hunters of the Lion burn its body to ashes, and sprinkle a piece of flesh with the powder, and so kill the Lion by means of its ashes even—so fatal to it is this poison! The Lion, therefore, not without reason, hates the Leontophonus, and, after destroying its sight, kills it without inflicting a bite: the animal, on the other hand, sprinkles the Lion with its urine, being well aware that this, too, is fatal to it."

We have read, in the Romances of Chivalry, how that Guy, Earl of Warwick, having seen a Lion and a

[1] From Λεοντοφονος, the Lion Killer.

Dragon fighting, went to the assistance of the former, and, having killed its opponent, the Lion meekly trotted after him, and ever after, until its death, was his constant companion. How, in the absence of Sir Bevis of Hampton, two lions having killed the Steward Boniface, and his horse, laid their heads in the fair Josian's lap. The old romancists held that a lion would always respect a virgin, and Spenser has immortalised this in his character of Una. Most of us remember the story given by Aulus Gellius and Ælian, of Androcles, who earned a lion's gratitude by extracting a thorn from its paw, and Pliny gives similar instances:—

"Mentor, a native of Syracuse, was met in Syria by a lion, who rolled before him in a suppliant manner; though smitten with fear, and desirous to escape, the wild beast on every side opposed his flight, and licked his feet with a fawning air. Upon this, Mentor observed on the paw of the lion, a swelling and a wound; from which, after extracting a splinter, he relieved the creature's pain.

"In the same manner, too, Elpis, a native of Samos, on landing from a vessel on the coast of Africa, observed a lion near the beach, opening his mouth in a threatening manner; upon which he climbed a tree, in the hope of escaping, while, at the same time, he invoked the aid of Father Liber (*Bacchus*); for it is the appropriate time for invocations where there is no room left for hope. The wild beast did not pursue him when he fled, although he might easily have done so; but, lying down at the foot of the tree, by the open mouth which had caused so much terror, tried to excite his compassion. A bone, while he was devouring his food with too great avidity, had stuck fast between his teeth, and he was perishing with hunger; such being the punishment inflicted upon him by his own weapons, every now and then he would look up, and supplicate him, as it were, with mute entreaties. Elpis, not wishing to risk trusting himself to so formidable a beast, remained stationary for some time, more at last from astonishment than from fear. At length, however, he descended from the tree, and extracted the bone, the lion, in the meanwhile, extending his head, and aiding in the operation as far as it was necessary for him to do. The story goes on to say, that as long as the vessel remained off that coast, the lion shewed his sense of gratitude by bringing whatever he had chanced to procure in the chase."

The same author mentions two curious animals, the Leucrocotta, and the Eale, which are noticeable among other wonders:—"Æthiopia produces the lynx in abundance, and the sphinx, which has brown hair and two mammæ on the breast, as well as many monstrous kinds of a similar nature; horses with wings, and armed with horns, which are called pegasi: the Crocotta, an animal which looks as though it had been produced by the union of the wolf and the dog, for it can break anything with its teeth, and instantly, on swallowing it, it digests it with the stomach; monkeys, too, with black heads, the hair of the ass, and a voice quite unlike that of any other animal."

The Leucrocotta—The Eale—Cattle Feeding Backwards.

"There are oxen, too, like that of India, some with one horn, and others with three; the leucrocotta, a wild beast of extraordinary swiftness, the size of the wild ass, with the legs of a Stag, the neck, tail, and breast of a lion, the head of a badger, a cloven hoof, the mouth slit up as far as the ears, and one continuous bone instead of teeth; it is said, too, that this animal can imitate the human voice.

"Among the same people there is found an animal called the eale; it is the size of the river-horse, has the tail of the elephant, and is of a black or tawny colour. It has, also, the jaws of the wild boar and horns that are moveable, and more than a cubit in length, so that, in fighting, it can employ them alternately, and vary their position by presenting them directly, or obliquely, according as necessity may dictate."

The Eale, with its movable horns, is run hard by the Cattle of the Lotophagi, which are thus described by Herodotus:—"From the Augilæ at the end of another ten days' journey is another hill of salt and water, and many fruit-bearing palm trees, as also in other places; and men inhabit it, who are called Gavamantes, a very powerful nation; they lay earth upon the salt, and then sow their ground. From these to the Lotophagi, the shortest route is a journey of thirty days: amongst them the kine that feed backwards are met with; they feed backwards for this reason. They have horns that are bent forward, therefore they draw back as they feed; for they are unable to go forward, because their horns would stick in the ground. They differ from other kine in no other respect than this, except that their hide is thicker and harder."

Animal Medicine.

We have already seen some of the wonderfully curative properties of animals—let us learn something of their own medical attainments—as described by Pliny. "The hippopotamus has even been our instructor in one of the operations of medicine. When the animal has become too bulky, by continued overfeeding, it goes down to the banks of the river, and examines the reeds which have been newly cut; as soon as it has found a stump that is very sharp, it presses its body against it, and so wounds one of the veins in the thigh; and by the flow of blood thus produced, the body, which would otherwise have fallen into a morbid state, is relieved; after which, it covers up the wound with mud.

"The bird, also, which is called the Ibis, a native of the same country of Egypt, has shewn us some things of a similar nature. By means of its hooked beak, it laves the body through that part by which it is especially necessary for health, that the residuous food should be discharged. Nor, indeed, are these the only inventions which have been borrowed from animals to prove of use to man. The power of the herb *dittany*, in extracting arrows, was first disclosed to us by stags that had been struck by that weapon; the weapon being discharged on their feeding upon this plant. The same animals, too, when they happen to have been wounded by the *phalangium*, a species of spider, or by any insect of a similar nature, cure themselves by eating crabs. One of the very best remedies for the bite of the serpent, is the plant with which lizards treat their wounds when injured in fighting with each other. The swallow has shown us that the *chelidonia* is very serviceable to the sight, by the fact of its employing it for the cure of its young, when their eyes are affected. The tortoise recruits its powers of effectually resisting serpents by eating the plant which is known as *cunile bubula*; and the weasel feeds on *rue*, when it fights with the serpent in pursuit of mice. The Stork cures itself of its diseases, with *wild marjoram*, and the wild boar with *ivy*, as also by eating *crabs*, and, more particularly, those that have been thrown up by the sea.

"The snake, when the membrane which covers its body, has been contracted by the cold of winter, throws it off in the spring, by the aid of the juices of *fennel*, and thus becomes sleek and youthful in appearance. First of all it disengages the head, and then it takes no less than a day and a night in working itself out, and divesting itself of the membrane in which it has been enclosed. The same animal, too, on finding its sight weakened during its winter retreat, anoints and refreshes its eyes by rubbing itself on the plant called *fennel*, or *marathrum*; but, if any of the scales are slow in coming off, it rubs itself against the thorns of the *juniper*. The dragon relieves the nausea which affects it in spring, with the juices of the *lettuce*. The barbarous nations go to hunt the panther, provided with meat that has been rubbed with *Aconite*, which is a poison. Immediately on eating it, compression of the throat overtakes them, from which circumstance it is, that the plant has received the name of *pardalianches* (*pard-strangler*). The animal, however, has found an antidote against this poison in human excrements; besides which, it is so eager to get at them, that the shepherds purposely suspend them in a vessel, placed so high, that the animal cannot reach them, even by leaping, when it endeavours to get at them; accordingly, it continues to leap, until it has quite exhausted itself, and at last expires: otherwise, it is so tenacious of life that it will continue to fight, long after its intestines have been dragged out of its body.

CURIOUS CREATURES.

"When an elephant has happened to devour a chameleon, which is of the same colour with the herbage, it counteracts this poison by means of the *wild olive*. Bears, when they have eaten of the fruit of the *Mandrake*, lick up numbers of Ants. The Stag counteracts the effect of poisonous plants by eating the *artichoke*. Wood pigeons, jackdaws, blackbirds, and partridges, purge themselves once a year by eating *bay* leaves; pigeons, turtle-doves, and poultry, with *wall pellitory*, or *helxine*; ducks, geese, and other aquatic birds of a similar nature, with the *bulrush*. The raven, when it has killed a chameleon, a contest in which even the conqueror suffers, counteracts the poison by means of laurel."

THE SU.

Topsell mentions a fearful beast called the Su. "There is a region in the new-found world, called *Gigantes*, and the inhabitants thereof, are called *Patagones;* now, because their country is cold, being far in the South, they cloath themselves with the skins of a beast called in their owne toong *Su*, for by reason that this beast liveth for the most part neere the waters, therefore they cal it by the name of *Su*, which signifieth water. The true image thereof, as it was taken by *Thenestus*, I have heere inserted, for it is of a very deformed shape, and monstrous presence, a great ravener, and an untamable wilde beast.

"When the hunters that desire her skinne, set upon her, she flyeth very swift, carrying her yong ones upon her back, and covering them with her broad taile; now, for so much as no dogge or man dareth to approach neere unto her, (because such is the wrath thereof, that in the pursuit she killeth all that commeth near her :) The hunters digge severall pittes or great holes in the earth, which they cover with boughes, sticks, and earth, so weakly, that if the beast chance at any time to come upon it, she, and her young ones fall down into the pit, and are taken.

"This cruell, untamable, impatient, violent, ravening, and bloody beast, perceiving that her natural strength cannot deliver her from the wit and policy of men, her hunters, (for being inclosed, she can never get out againe) the hunters being at hand to watch her down-

fall, and worke her overthrowe, first of all to save her young ones from taking and taming, she destroyeth them all with her own teeth; for there was never any of them taken alive, and when she seeth the hunters come about her, she roareth, cryeth, howleth, brayeth, and uttereth such a fearefull, noysome, and terrible clamor, that the men which watch to kill her, are not thereby a little amazed; but, at last, being animated, because there can be no resistance, they approach, and with their darts and speares, wound her to death, and then take off her skin, and leave the Carcasse in the earth. And this is all that I finde recorded of this most strange beast."

THE LAMB-TREE.

As a change from this awful animal, let us examine the *Planta Tartarica Borometz*—which was so graphically delineated by Joannes Zahn in 1696. Although this is by no means the first picture of it, yet it is the best of any I have seen.

A most interesting book[1] on the "Vegetable Lamb of Tartary" has been written by the late Henry Lee, Esq., at one time Naturalist of the Brighton Aquarium, and I am much indebted to it for matter on the subject, which I could not otherwise have obtained.

The word *Borometz* is supposed to be derived from a Tartar word signifying a lamb, and this plant-animal was thoroughly believed in, many centuries ago—but there seem to have been two distinct varieties of plant, that on which little lambs were found in pods, and that as represented by Zahn, with a living lamb attached by its navel to a short stem. This stalk was flexible, and allowed the lamb to graze, within

[1] Written to prove that this plant was the Cotton-plant.

its limits; but when it had consumed all the grass within its reach, or if the stalk was severed, it died. This lamb was said to have the actual body, blood, and bones of a young sheep, and wolves were very fond of it—but, luckily for the lamb-tree, these were the only carnivorous animals that would attack it.

In his "Histoire Admirable des Plantes" (1605) Claude Duret, of Moulins, treats of the Borometz, and says: " I remember to have read some time ago, in a very ancient Hebrew book entitled in Latin the *Talmud Ierosolimitanum*, and written by a Jewish Rabbi Jochanan, assisted by others, in the year of Salvation 436, that a certain personage named Moses Chusensis (he being a native of Ethiopia) affirmed, on the authority of Rabbi Simeon, that there was a certain country of the earth which bore a zoophyte, or plant-animal, called in the Hebrew *Jeduah*. It was in form like a lamb, and from its navel, grew a stem or root by which this Zoophyte, or plant-animal, was fixed attached, like a gourd, to the soil below the surface of the ground, and, according to the length of its stem or root, it devoured all the herbage which it was able to reach within the circle of its tether. The hunters who went in search of this creature were unable to capture, or remove it, until they had succeeded in cutting the stem by well-aimed arrows, or darts, when the animal immediately fell prostrate to the earth, and died. Its bones being placed with certain ceremonies and incantations in the mouth of one desiring to foretell the future, he was instantly seized with a spirit of divination, and endowed with the gift of prophecy."

Mr. Lee then says: "As I was unable to find in the Latin translation of the Talmud of Jerusalem, the passage mentioned by Claude Duret, and was anxious to ascertain whether any reference to this curious legend existed in the Talmudical books, I sought the assistance of learned members of the Jewish community, and, amongst them, of the Rev. Dr. Hermann Adler, Chief Rabbi Delegate of the United Congregations of the British Empire. He most kindly interested himself in the matter, and wrote to me as follows: 'It affords me much gratification to give you the information you desire on the Borametz. In the Mishna *Kilaim*, chap. viii. § 5 (a portion of the Talmud), the passage occurs: "Creatures called *Adne Hasadeh* (literally 'lords of the field') are regarded as beasts." There is a variant reading, *Abne Hasadeh* (stones of the field). A commentator, Rabbi Simeon, of Sens (died about 1235), writes as follows, on this passage: 'It is stated in the Jerusalem Talmud that this is a human being of the mountains: it lives by means of its navel: if its navel be cut, it cannot live. I have heard in the name of Rabbi Meir, the son of Kallonymos of Speyer, that this is the animal called *Jeduah*. This is the *Jedoui* mentioned in Scripture (lit. *wizard*, Lev. xix. 31); with its bones witchcraft is practised. A kind of large stem issues from a root in the earth on which this animal, called *Jadua*, grows, just as gourds and melons. Only the *Jadua* has, in all respects, a human shape, in face, body, hands, and feet. By its navel it is joined to the stem that issues from the root. No creature can approach within the tether of the stem, for it seizes and kills them. Within the tether of the stem it devours the herbage all around. When they want to capture it, no man dares approach it, but they tear at the stem until it is ruptured, whereupon the animal dies.' Another commentator, Rabbi Obadja, of Berbinoro, gives the same explanation, only substituting 'They aim arrows at the stem until it is ruptured,' &c.

"The author of an ancient Hebrew work, *Maase Tobia* (Venice, 1705), gives an interesting description of this animal. In Part IV. c. 10, page 786, he mentions the Borametz found in Great Tartary. He repeats the description of Rabbi Simeon, and adds, that he has found, in 'A New Work on Geography,' namely, that 'the Africans (*sic*) in Great Tartary, in the province of Sambulala, are enriched by means of seeds, like the seeds of gourds, only shorter in size, which grow and blossom like a stem to the navel of an animal which is called *Borametz* in their language, i.e. *lamb*, on account of its resembling a lamb in all its limbs, from head to foot; its hoofs are cloven, its skin is soft, its wool is adapted for clothing, but it has no horns, only the hairs of its head, which grow, and are intertwined like horns. Its height is half a cubit and more. According to those who speak of this wondrous thing, its taste is like the flesh of fish, its blood as sweet as honey, and it lives as long as there is herbage within reach of the stem, from which it derives its life. If the herbage is destroyed or perishes, the animal also dies away. It has rest from all beasts and birds of prey, except the wolf, which seeks to destroy it.' The author concludes by expressing his belief that this account of the animal having the shape of a lamb is more likely to be true than it is of human form."

As I have said, there are several delineations of this Borametz or Borometz, but there is one, a frontispiece to the 1656 edition of the *Paridisi in Sole—Paradisus Terrestris*, of John Parkinson, Apothecary of London, in which, together with Adam and Eve, the *lamb-tree* is shown as flourishing in the Garden of Eden; and Du Bartas, in

"His *divine* WEEKES *And* WORKES" in his poem of Eden, (the first day of the second week), makes Adam to take a tour of Eden, and describes his wonder at what he sees, especially at the "lamb-plant."

> "Musing, anon through crooked Walks he wanders,
> Round-winding rings, and intricate Meanders,
> Fals-guiding paths, doubtfull beguiling strays,
> And right-wrong errors of an end-less Maze:
> Not simply hedged with a single border
> Of *Rosemary*, cut-out with curious order,
> In *Satyrs, Centaurs, Whales,* and *half-men-Horses,*
> And thousand other counterfaited corses;
> But with true Beasts, fast in the ground still sticking,
> Feeding on grass, and th' airy moisture licking:
> Such as those *Bonarets*, in *Scythia* bred
> Of slender seeds, and with green fodder fed;
> Although their bodies, noses, mouthes and eys,
> Of new-yean'd Lambs have full the form and guise;
> And should be very Lambs, save that (for foot)
> Within the ground they fix a living root,
> Which at their navell growes, and dies that day
> That they have brouz'd the neighbour grass away.
>   O wondrous vertue of God onely good!
> The Beast hath root, the Plant hath flesh and blood
> The nimble Plant can turn it to and fro;
> The nummed Beast can neither stir nor go:
> The Plant is leaf-less, branch-less, void of fruit;
> The Beast is lust-less, sex-less, fire-less, mute;
> The Plant with Plants his hungry panch doth feed;
> Th' admired Beast is sowen a slender seed."

Of the other kind of "lamb-tree," that which bears lambs in pods, we have an account, in Sir John Maundeville's Travels. "Whoso goeth from Cathay to Inde, the high and the low, he shal go through a Kingdom that men call Cadissen, and it is a great lande, there groweth a manner of fruite as it were gourdes, and when it is ripe men cut it a sonder, and men fynde therein a beast as it were of fleshe and bone and bloud, as it were a lyttle lambe without wolle, and men eate the beaste and fruite also, and sure it seemeth very strange."

And in the " Journall of Frier Odoricus," which I have incorporated in my edition of " The Voiage and Travayle of Syr John Maundeville, Knight," he says: " I was informed also by certaine credible persons of another miraculous thing, namely, that in a certaine Kingdome of the sayd Can, wherein stand the mountains called Kapsei (the Kingdomes name is Kalor) there groweth great Gourds or Pompions, (*pumpkins*) which being ripe, doe open at the tops, and within them is found a little beast like unto a yong lambe."

THE CHIMÆRA.

Aldrovandus gives us the accompanying illustration of a Chimæra, a fabulous Classical monster, said to pos-

CURIOUS CREATURES. 171

sess three heads, those of a lion, a goat, and a dragon. It used so to be pictorially treated, but in more modern times as Aldrovandus represents. The mountain *Chimæra*, now called Yanar, is in ancient Lycia, in Asia Minor, and was a burning mountain, which, according to Spratt, is caused by a stream of inflammable gas, issuing from a crevice. This monster is easily explained, if we can believe Servius, the Commentator of Virgil, who says that flames issue from the top of the mountain, and that there are lions in the vicinity; the middle part abounds in goats, and the lower part with serpents.

THE HARPY AND SIREN.

The conjunction of the human form with birds is very easy, wings being fitted to it, as in the case of angels—and as applied to beasts, this treatment is very ancient, *vide* the winged bulls of Assyria, and the classical Pegasus, or winged horse. With birds, the best form in which it is treated in Mythology is the Harpy. This is taken from Aldrovandus, and fully illustrates the mixture of bird and woman, described by Shakespeare in *Pericles* (iv. 3) :—

> "*Cleon.*   Thou'rt like the harpy,
> Which to betray, dost, with thine angel's face,
> Seize with thine eagle's talons."

Then, also, we have the Siren, shown by this illustration, taken from Pompeii. These Sea Nymphs were like the Harpies, depicted as a compound of bird and woman.

Like them also, there were three of them; but, unlike them, they had such lovely voices, and were so beautiful, that they lured seamen to their destruction, they having no power to combat the allurements of the Sirens; whilst the Harpies emitted an infectious smell, and spoiled whatever they touched, with their filth, and excrements.

Licetus, writing in 1634, and Zahn, in 1696, give the accompanying picture of a monster born at Ravenna in

1511 or 1512. It had a horn on the top of its head, two wings, was without arms, and only one leg like that of a bird of prey. It had an eye in its knee, and was of both sexes. It had the face and body of a man, except in the lower part, which was covered with feathers.

Marcellus Palonius Romanus made some Latin verses upon this prodigy, which may be thus rendered into English :—

> A Monster strange in fable, and deform
> Still more in fact; sailing with swiftest wing,
> He threatens double slaughter, and converts
> To thy fell ruin, flames of living fire.
> Of double sex, it spares no sex, alike
> With kindred blood it fills th' Æmathian plain ;
> Its corpses strew alike both street and sea.
> There hoary Thetis and the Nereids
> Swim shudd'ring through the waves, while floating wide
> The fish replete on human bodies——. Such,
> Ravenna, was the Monster which foretold
> Thy fall, which brings thee now such bitter woe,
> Tho' boasting in thy image triumph-crowned.

The Barnacle Goose.

Of all extraordinary beliefs, that in the Barnacle Goose, which obtained credence from the eleventh to the seventeenth centuries, is as wonderful as any. The then accepted fact that the Barnacle Goose was generated on trees, and dropped alive in the water, dates back a hundred years before Gerald de Barri. Otherwise Giraldus Cambrensis wrote in 1187, about these birds, the following being a translation :—

"There are here many birds which are called Bernacæ, which nature produces in a manner contrary to nature, and very wonderful. They are like marsh-geese, but smaller. They are produced from fir timber tossed about at sea, and are at first like geese upon it. Afterwards they hang down by their beaks, as if from a seaweed attached to the wood, and are enclosed in shells that they may grow the more freely. Having thus, in course of time, been clothed with a strong covering of feathers, they either fall into the water, or seek their liberty in the air by flight. The embryo geese derive their growth and nutriment from the moisture of the wood or of the sea, in a secret and most marvellous manner. I have seen with my own eyes more than a thousand minute bodies of these birds hanging from one piece of timber on the shore, enclosed in shells, and already formed. The eggs are not impregnated *in coitu*, like those of other birds, nor does the bird sit upon its eggs to hatch them, and in no corner of the world have they been known to build a nest. Hence the bishops and clergy in some parts of Ireland are in the habit of partaking of these birds, on fast days, without scruple. But in doing so they are led into sin. For, if any one were to eat of the leg of our first parent, although he (Adam) was not born of flesh, that person could not be adjudged innocent of eating flesh."

We see here, that Giraldus speaks of these barnacles being developed on wreckage in the sea, but does not mention their growing upon trees, which was the commoner belief. I have quoted both Sir John Maundeville, and Odoricus, about the lamb-tree, which neither seem to consider very wonderful, for Sir John says: " Neverthelesse I sayd to them that I held y^t for no marvayle, for I sayd that in my countrey are trees y^t beare fruit, y^t become byrds flying, and they are good to eate, and that that falleth on the water, liveth, and that that falleth on earth, dyeth, and they marvailed much thereat." And the Friar, in continuation of his story of the *Borometz*, says: "Even as I my selfe have heard reported that there stand certaine trees upon the

shore of the Irish Sea, bearing fruit like unto a gourd, which at a certaine time of the yeere doe fall into the water, and become birds called Bernacles, and this is most true."

Olaus Magnus, in speaking of the breeding of Ducks in Scotland, says: "Moreover, another *Scotch* Historian, who diligently sets down the secret of things, saith that in the *Orcades,* (*the Orkneys*) Ducks breed of a certain

Fruit falling in the Sea; and these shortly after, get wings, and fly to the tame or wild ducks." And, whilst discoursing on Geese, he affirms that "some breed from Trees, as I said of Scotland Ducks in the former Chapter." Sebastian Müenster, from whom I have taken the preceding illustration, says in his *Cosmographia Universalis:* —" In Scotland there are trees which produce fruit, conglomerated of their leaves; and this fruit, when, in due time, it falls into the water beneath it, is endowed with new life, and is converted into a living bird, which they call the 'tree goose.' This tree grows in the Island of Pomonia, which is not far from Scotland, towards the North. Several old Cosmographers, especially Saxo Grammaticus, mention the tree, and it must not be regarded as fictitious, as some new writers suppose."

In Camden's "Britannia" (translated by Edmund Gibson, Bishop of London) he says, speaking of Buchan:— "It is hardly worth while to mention the clayks, a sort of geese; which are believed by some, (with great admiration) to grow upon the trees on this coast and in other places, and, when they are ripe, to fall down into the sea; because neither their nests nor eggs can anywhere be found. But they who saw the ship, in which Sir Francis Drake sailed round the world, when it was laid up in the river Thames, could testify, that little birds breed in the old rotten keels of ships; since a great number of such, without life and feathers, stuck close to the outside of the keel of that ship; yet I should think, that the generation of these birds was not from the logs of wood, but from the sea, termed by the poets 'the parent of all things.'"

In "Purchas, his Pilgrimage," is the voyage of Gerat de Veer to China, &c., in 1569—and he speaks of the

Barnacle goose thus :—"Those geese were of a perfit red colour, such as come to Holland about Weiringen, and every yeere are there taken in abundance, but till this time, it was never knowne where they hatcht their egges, so that some men have taken upon them to write

that they sit upon trees in Scotland, that hang over the water, and such eggs that fall from them downe into the water, become young geese, and swim there out of the water: but those that fall upon the land, burst asunder, and are lost; but that is now found to be contrary, that no man could tell where they breed their egges, for that no man that ever wee knew, had ever beene under 80°; nor that land under 80° was never set downe in any card, much lesse the red geese that breede therein." He and his sailors declared that they had seen these birds sitting on their eggs, and hatching them, on the coasts of Nova Zembla.

Du Bartas thus mentions this goose :—

> "So, slowe Boötes underneath him sees,
> In th' ycie iles, those goslings hatcht of trees ;
> Whose fruitfull leaves, falling into the water,
> Are turned, (they say) to living fowls soon after.
> So, rotten sides of broken ships do change
> To barnacles ; O transformation strange !
> 'Twas first a green tree, then a gallant hull,
> Lately a mushroom, now a flying gull."

I could multiply quotations on this subject. Gesner and every other naturalist believed in the curious birth of the Barnacle goose—and so even did Aldrovandus, writing at the close of the seventeenth century, for from him I take this illustration. But enough has been said upon the subject.

Remarkable Egg.

No wonder that a credulous age, which could see nothing extraordinary in the Barnacle goose, could also, metaphorically, swallow such an egg, as Licetus, first of all, and Aldrovandus, after him, gives us in the accompanying true picture. The latter says that a goose's egg was found in France, (he leaves a liberal margin for locality,) which on being broken appeared exactly as in the picture. Comment thereon is useless.

Moon Woman.

One would have imagined that this Egg would be sufficient to test the credulity of most people, but Aldrovandus was equal to the occasion, and he gives us a "Moon Woman," who lays eggs, sits upon them, and

hatches Giants; and he gives this on the authority of Lycosthenes and Ravisius Textor.

The Griffin.

There always has been a tradition of birds being existent, of far greater size than those usually visible.

The Maoris aver that at times they still hear the gigantic Moa in the scrub—and, even, if extinct, we know, by the state of the bones found, that its extinction must have been of comparatively recent date. But no one credits the Moa with the power of flight, whilst the Griffin, which must not be confounded with the gold-loving Arimaspian Gryphon, was a noble bird. Mandeville knew him :—" In this land (*Bactria*) are many gryffons, more than in other places, and some

say they have the body before as an Egle, and behinde as a Lyon, and it is trouth, for they be made so; but the Griffen hath a body greater than viii Lyons, and stall worthier (*stouter, braver*) than a hundred Egles. For certainly he wyl beare to his nest flying, a horse and a man upon his back, or two Oxen yoked togither as they go at plowgh, for he hath longe nayles on hys fete, as great as it were hornes of Oxen, and of those they make Cups there to drynke of, and of his rybes they make bowes to shoote with."

Olaus Magnus says they live in the far Northern

mountains, that they prey upon horses and men, and that of their nails drinking-cups were made, as large as ostrich eggs. These enormous birds correspond in many points to the Eastern Ruc or Rukh, or the Rok of the "Arabian Nights," of whose mighty powers of flight Sindbad took advantage.

Ser Marco Polo, speaking of Madagascar, says:—"'Tis said that in those other Islands to the south, which the ships are unable to visit because this strong current prevents their return, is found the bird *Gryphon*, which appears there at certain seasons. The description given of it is, however, entirely different from what our stories and pictures make it. For persons who had been there and had seen it, told Messer Marco Polo that it was for all the world like an eagle, but one indeed of enormous size; so big in fact, that its wings covered an extent of 30 paces, and its quills were 12 paces long, and thick in proportion. And it is so strong that it will seize an Elephant in its talons, and carry him high into the air, and drop him so that he is smashed to pieces: having so killed him, the bird gryphon swoops down on him, and eats him at leisure. The people of those isles call the bird *Ruc*, and it has no other name. So I wot not if this be the real gryphon, or if there be another manner of bird as great. But this I can tell you for certain, that they are not half lion and half bird, as our stories do relate; but, enormous as they be, they are fashioned just like an eagle.

"The Great Kaan sent to those parts to enquire about these curious matters, and the story was told by those who went thither. He also sent to procure the release of an envoy of his who had been despatched thither, and had been detained; so both those envoys had many wonderful things to tell the Great Kaan about those strange islands, and about the birds I have mentioned. They brought (as I heard) to the Great Kaan, a feather of the said Ruc, which was stated to measure 90 Spans, whilst the quill part was two palms in circumference, a marvellous object! The Great Kaan was delighted with it, and gave great presents to those who brought it."

This quill seems rather large; other travellers, however, perhaps not so truthful as Ser Marco, speak of these enormous quills. The Moa of New Zealand (*Dinornis giganteus*) is supposed to have been the largest bird in Creation—and next to that is the *Æpyornis maximus*—whose bones and egg have been found in Madagascar. An egg is in the British Museum, and it has a liquid capacity of 2.35 gallons, but, alas, for the quill story—this bird was wingless.

The Condor has been put forward as the real and veritable Ruc, but no living specimens will compare with this bird as it has been described—especially if we take the picture of it in Lane's "Arabian Nights," where it is represented as taking up *three* elephants, one in its beak, and one in each of its claws.

The Japanese have a legend of a great bird which carried off men—and there is a very graphic picture now on view at the White Wing of the British Museum, where one of these birds, having seized a man, frightens, very naturally, the whole community.

The Phœnix.

Pliny says of the Phœnix:—"Æthiopia and India, more especially produce birds of diversified plumage, and such as quite surpass all description. In the front rank of these is the Phœnix, that famous bird of Arabia; though I am not sure that its existence is not a fable.

"It is said that there is only one in existence in the whole world, and that that one has not been seen very often. We are told that this bird is of the size of an eagle, and has a brilliant golden plumage around the neck, whilst the rest of the body is a purple colour; except the tail, which is azure, with long feathers intermingled, of a roseate hue; the throat is adorned with a crest, and the head with a tuft of feathers. The first Roman who described this bird, and who has done so with great exactness, was the Senator Manilius, so famous for his learning; which he owed, too, to the instructions of no teacher. He tells us that no person has ever seen this bird eat, that in Arabia it is looked upon as sacred to the Sun; that it lives five hundred and forty years. That when it is old it builds a nest of Cassia and sprigs of incense, which it fills with perfumes, and then lays its body down upon them to die: that from its bones and marrow there springs at first a sort of small worm, which, in time, changes into a little bird; that the first thing it does is to perform the obsequies of its predecessor, and to carry the nest entire to the City of the Sun near Panchaia, and there deposit it upon the altar of that divinity.

"The same Manilius states also, that the revolution of the great year is completed with the life of this bird, and that then a new cycle comes round again with the same characteristics as the former one, in the seasons and the appearance of the stars; and he says that this begins about midday of the day in which the Sun enters the sign of Aries. He also tells us that when he wrote to the above effect, in the consulship of P. Licinius, and Cneius Cornelius, (B.C. 96) it was the two hundred and fifteenth year of the said revolution. Cornelius Valerianus says that the Phœnix took its flight from Arabia into Egypt in the Consulship of Q. Plautius and Sextus Papinius, (A.D. 36). This bird was brought to Rome in the Censorship of the Emperor Claudius, being the year from the building of the City, 800, (A.D. 47) and it was exposed to public view in the Comitium. This fact is attested by the public Annals, but there is no one that doubts that it was a fictitious Phœnix."

Cuvier seems to think that the bird described above was a Golden Pheasant, brought from the interior of Asia—at a time when these birds were unknown to civilised Europe.

Du Bartas, in his metrical account of the Creation, mentions this winged prodigy:—

> " The Heav'nly Phœnix first began to frame
> The earthly *Phœnix*, and adorn'd the same
> With such a Plume, that Phœbus, circuiting
> From *Fez* to *Cairo*, sees no fairer thing:
> Such form, such feathers, and such Fate he gave her
> That fruitfull Nature breedeth nothing braver:
> Two sparkling eyes; upon her crown, a crest
> Of starrie Sprigs (more splendent than the rest)
> A goulden doun about her dainty neck,
> Her brest deep purple, and a scarlet back,
> Her wings and train of feathers (mixed fine)
> Of orient azure and incarnadine.
> He did appoint her Fate to be her Pheer,
> And Death's cold kisses to restore her heer
> Her life again, which never shall expire
> Untill (as she) the World consume in fire.
> For, having passed under divers Climes,
> A thousand Winters, and a thousand Primes;
> Worn out with yeers, wishing her endless end,

To shining flames she doth her life commend,
Dies to revive, and goes into her Grave
To rise againe more beautifull and brave.
With Incense, Cassia, Spiknard, Myrrh, and Balm,
By break of Day shee builds (in narrow room)
Her Urn, her Nest, her Cradle, and her Toomb;
Where, while she sits all gladly-sad expecting
Some flame (against her fragrant heap reflecting)
To burn her sacred bones to seedfull cinders,
(Wherein, her age, but not her life, she renders.)

. . . . . .

And *Sol* himself, glancing his goulden eyes
On th' odoriferous Couch wherein she lies,
Kindles the spice, and by degrees consumes
Th' immortall *Phœnix*, both her flesh and plumes.
But instantly, out of her ashes springs
A Worm, an Egg then, then a Bird with wings,
Just like the first, (rather the same indeed)
Which (re-ingendred of its selfly seed)
By nobly dying, a new Date begins,
And where she loseth, there her life she wins :
Endless by'r End, eternall by her Toomb ;
While, by a prosperous Death, she doth becom
(Among the cinders of her sacred Fire)
Her own selfs Heir, Nurse, Nurseling, Dam and Sire."

The Swallow.

"And is the swallow gone?
Who beheld it?
Which way sailed it?
Farewell bade it none?"

(*W. Smith, Country book.*)

Olaus Magnus answered this question, according to his lights, and when, discoursing on the Migration of Swallows he says :—" Though many Writers of Natural Histories have written that Swallows change their stations; that is, when cold Winter begins to come, they fly to hotter Climats; yet oft-times, in the Northern Countries, Swallows are drawn forth, by chance by Fishermen, like a lump cleaving together, where they went amongst the Reeds, after the beginning of Autumn, and there fasten themselves bill to bill, wing to wing, feet to feet. For it is observed, that they, about that time ending their most sweet note, (?) do so descend, and they fly out peaceably after the beginning of the Spring, and come to their old Nests, or else they build new ones by their natural care. Now that lump being drawn forth by

ignorant young men (for the old Fishermen that are acquainted with it, put it in again) is carryed and laid on the Sea Shore, and by the heat of the Sun, the Lump is dissolved, and the Swallows begin to fly, but they last but a short time because they were not set at liberty by being taken so soon, but they were made captive by it. It hapneth also in the Spring, when they return freely, and come to their old Nests, or make new ones, if a very cold Winter come upon them, and much snow fall, they will all dye; that all that Summer you shall see none of them upon the Houses, or Banks, or Rivers; but a very few that came later out of the Waters, or from other Parts, which by Nature come flying thither, to repair their Issue. Winter being fully ended in *May;* For Husband-Men, from their Nests, built higher or lower, take their Prognostications, whether they shall sowe in Valleys, or Mountains or Hills, according as the Rain shall increase or diminish. Also the Inhabitants hold it an ill sign, if the Swallows refuse to build upon their houses; for they fear those House-tops are ready to fall."

This is proper, and good, and what we might expect from Olaus Magnus; but it is somewhat singular to see, printed in *Notes and Queries* for October 22, 1864, the following:—

"The Duke de R——— related to me, a few days ago, that in Sweden, the swallows, as soon as the winter begins to approach, plunge themselves into the lakes, where they remain asleep and hidden under the ice till the return of the summer; when, revived by the new warmth, they come out from the water, and fly away as formerly. While the lakes are frozen, if somebody will break the ice in those parts where it appears darker than in the rest, he will find masses of swallows—cold, asleep, and half dead; which, by taking out of their retreat, and warming, he will see gradually to vivify again and fly.

"In other countries they retire very often to the Caverns, under the rocks. As many of these exist between the City of Caen, and the Sea, on the banks of the river Orne, there are found sometimes, during the winter, piles of swallows suspended in these vaults, like bundles of grapes. I witnessed the same thing, myself, in Italy; where, as well as in France, it is considered (as I have heard) very lucky by the inhabitants when swallows build nests on their habitations. . . . . *Rhodocanakis*."

Of course, these stories of curious hybernation were pooh-poohed, although it could not be denied that the subaqueous hybernation of swallows is given in Goldsmith's "Animated Nature," and many other Natural Histories, which succeeded his.

The wintering of swallows in caverns, has another eye-witness in Edward Williams (*Iolo Morganwg*), who in his "Poems, Lyrics, and Pastorals," published 1794, says: —"About the year 1768, the author, with two or three more, found a great number of swallows in a torpid state, clinging in clusters to each other by their bills, in a cave of the sea-cliffs near Dunraven Castle, in the County of Glamorgan. They revived after they had been some hours in a warm room, but died a day or two after, though all possible care had been taken of them."

The Martlet, and Footless Birds.

Of the Martin, or, as in Heraldry it is written, *Martlet*, Guillim thus writes:—" The Martlet, or Martinet, saith Bekenhawh, hath Legs so exceeding short, that they can by no means go : (*walk*) And thereupon, it seemeth, the *Grecians* do call them *Apodes, quasi sine pedibus;* not because they do want Feet, but because they have not such Use of their Feet, as other Birds have. And if perchance they fall upon the Ground, they cannot raise themselves upon their Feet, as others do, and prepare themselves to flight. For this Cause they are accustomed to make their Nests upon Rocks and other high places, from whence they may easily take their flight, by Means of the Support of the Air. Hereupon it came, that this Bird is painted in Arms without Feet: and for this Cause it is also given for a Difference of younger Brethren, to put them in mind to trust to their wings of Vertue and Merit, to raise themselves, and not to their Legs, having little Land to put their foot on."

The Alerion is a small bird of the eagle tribe, heraldically depicted as without beak or feet.

Butler in "Hudibras" writes—

> "Like a bird of paradise,
> Or herald's Martlet, has no legs,
> Nor hatches young ones, nor lays eggs."

The Bird of Paradise was unknown to the ancients, and one of the earliest notices of this bird is given in Magalhaen's voyage in 1521:—"The King of Bachian, one of the Molucca Islands, sent two dead birds preserved, which were of extraordinary beauty. In size they were not larger than the thrush: the head was small, with a long bill; the legs were of the thickness of a common quill, and a span in length; the tail resembled that of the thrush; they had no wings, but in the place where wings usually are, they had tufts of long feathers, of different colours; all the other feathers were dark. The inhabitants of the Moluccas had a tradition that this bird came from Paradise, and they call it *bolondinata*, which signifies the 'bird of God.'"

By-and-by, as trade increased, the skins of this bird were found to have a high market value, but the natives always brought them, when they came to trade, with their legs cut off. Thence sprang the absurd rumour that they had no legs, although in the early account just quoted, their legs are expressly mentioned. Linnæus called the emerald birds of Paradise *apoda* or legless; whilst Tavernier says that these birds getting drunk on nutmegs, fall helpless to the ground, and then the ants eat off their legs.

> "But note we now, towards the rich *Moluques*,
> Those passing strange and wondrous (birds) *Manueques*.
> (Wond'rous indeed, if Sea, or Earth, or Sky,
> Saw ever wonder swim, or goe, or fly)
> None knowes their Nest, none knowes the dam that breeds them;
> Foodless they live; for th' Aire alonely feeds them:
> Wingless they fly; and yet their flight extends,
> Till with their flight, their unknown live's-date ends."

Snow Birds.

But we must leave warm climes, and birds of Paradise, and speak of "Birds shut up under the Snow."

"There are in the Northern Countries Wood-Cocks,

like to pheasant for bigness, but their Tails are much shorter, and they are cole black all over their bodies, with some white feathers at the end of their Tails and

Wings. The Males have a red Comb standing upright; the Females have one that is low and large, and the colour is grey. These Birds are of an admirable Nature to endure huge Cold in the Woods, as the Ducks in the Waters. But when the Snow covers the Superficies of the Earth, like to Hills, all over, and for a long time presse down the boughs of the Trees with their weight, they eat certain Fruits of the Birch-Tree, called in *Italian* (*Gatulo*) like to a long Pear, and they swallow them whole, and that in so great quantity, and so greedily, that their throat is stuffed, and seems greater than all their body.

"Then they part their Companies, and thrust themselves all over into the snow, especially in *January*, *February* and *March*, when Snow and Whirlwinds, Storms, and grievous Tempests, descend from the Clouds. And when they are covered all over, that not one of them can be seen, lying all in heaps, for certain weeks they live, with meat collected in their throats, and cast forth, and resumed. The Hunter's Dogs cannot find them; yet by the Cunning of the crafty Hunters, it falls out, that when the Dogs err in their scent, they, by signs, will catch a number of living Birds, and will draw them forth to their great profit. But they must do that quickly; because when they hear the Dogs bark, they presently rise like Bees, and take up on the Wing, and fly aloft. But, if they perceive that the Snow will be greater, they devour the foresaid Fruit again, and take a new dwelling, and there they stay till the end of March: or, if the snow melt sooner, when the Sun goes out of *Aries;* for then the snow melting, by an instinct of Nature (as many other Birds) they rise out of their holes to lay Eggs, and produce young ones; and this in Mountains where bryars are, and thick Trees. Males and Females sit on the Eggs by turns, and both of them keep the Young, and chiefly the Male, that neither the Eagle nor Fox may catch them.

"These Birds fly in great sholes together, and they remain in high Trees, chiefly Birch-Trees; and they come not down, but for propagation, because they have food enough on the top of their Trees. And when Hunters or Countreymen, to whom those fields belong, see them fly all abroad, over the fields full of snow, they pitch up staves obliquely from the Earth, above the Snow, eight or ten foot high; and at the top of them, there hangs a snare, that moves with the least touch, and so they catch these Birds; because they, when they Couple, leap strangely, as Partridges do, and so they fall into these snares, and hang there. And when one seems to be caught in the Gin, the others fly to free her, and are caught in the like snare. There is also another way to catch them, namely with arrows and stalking-horses, that they may not suspect it. . . .

"There is also another kind of Birds called *Bonosa*, whose flesh is outwardly black, inwardly white: they are as delicate good meat as Partridges, yet as great as Pheasants. At the time of Propagation, the Male runs with open mouth till he foam; then the Female runs and receives the same; and from thence she seems to conceive, and bring forth eggs, and to produce her young."

The Swan.

The ancient fable so dear, even to modern poets, that Swans sing before they die—was not altogether believed even in classical times, as saith Pliny:—"It is stated that at the moment of the swan's death, it gives utterance to a mournful song; but this is an error, in my opinion; at least, I have tested the truth of the story on several occasions." That some swans have a kind of voice, and can change a note or two, no one who has met with a flock or two of "hoopers," or wild swans, can deny.

Olaus Magnus relates the fable—and quotes Plato, that the swan sings at its death, not from sorrow, but out of joy, at finishing its life. He also gives us a graphic illustration of how swans may be caught by playing to them on a lute or other stringed instrument, and also that they

were to be caught by men (playing music) with stalking-horses, in the shape of oxen, or horses; and, in another page, he says, that not far from London, the Metropolis of England, on the River Thames, may be found more than a thousand domesticated swans.

The Alle, Alle.

"There is also in this Lake (*the White Lake*) a kind of bird, very frequent; and in other Coasts of the *Bothnick*

CURIOUS CREATURES. 195

and *Swedish* Sea, that cries incessantly all the Summer, *Alle, Alle*, therefore they are called all over, by the Inhabitants, *Alle, Alle*. For in that Lake such a multitude of great birds is found, (as I said before) by reason of the fresh Waters that spring from hot springs, that they seem to cover all the shores and rivers, especially Sea-Crows, or Cormorants, Coots, More Hens, two sorts of Ducks, Swans, and infinite smaller Water Birds. These Crows, and other devouring birds, the hunters can easily take, because they fly slowly, and not above two or four

Cubits above the Water: thus they do it on the narrow Rocks, as in the Gates of Islands, on the Banks of them, they hang black nets, or dyed of a Watry Colour upon Spears; and these, with Pulleys, will quickly slip up or down, that in great Sholes they catch the Birds that fly thither by letting the Nets fall upon them: and this is necessary, because those Birds fly so slowly, and right forward; so that few escape. Also, sometimes Ducks, and other Birds are taken in these Nets. Wherefore these black, or slow Birds, whether they swim or fly, are always crying *Alle, Alle*, which in Latine signifies *All, All*,

(*Omnes*) and so they do when they are caught in the Nets: and this voyce the cunning Fowler interprets thus, that he hath not, as yet, all of them in his Nets; nor ever shall have, though he had six hundred Nets."

THE HOOPOE AND LAPWING.

Whether the following bird is meant for the Hoopoe, or the Lapwing, I know not. The Latin version has "De Upupis," which clearly means Hoopoes—and the translation says, "Of the Whoups or Lapwings"—I follow the latter. "*Lapwings*, when at a set time they come to the Northern Countries from other parts, they foreshew

the nearnesse of the Spring coming on. It is a Bird that is full of crying and lamentation, to preserve her Eggs, or young. By importunate crying, she shews that Foxes lye hid in the grasse; and so she cries out in all places, to drive away dogs and other Beasts. They fight with Swallows, Pies, and Jackdaws.

"On Hillocks, in Lakes, she lays her Eggs, and hatcheth

her young ones. Made tame she will cleane a house of Flyes, and catch Mice. She foreshews Rain when she cries; which also Field Scorpions do, called Mares, Cuckows; who by flying overthwart, and crying loudly, foreshew Rain at hand; also the larger Scorpions, with huge long snouts, fore signifie Rain; so do Woodpeckers. There is a Bird also called Rayn, as big as a Partridge that hath Feathers of divers colours, of a yellow, white, and black colour: This is supposed to live upon nothing but Ayr, though she be fat, nothing is found in her belly. The Fowlers hunt her with long poles, which they cast high in the Ayr to fright her, so that they may catch the Bird flying down."

The Ostrich.

Modern observation, and especially Ostrich farming, has thoroughly exploded the old errors respecting this bird. We believe in its powers of *swallowing* anything not too large, but not in its *digesting* everything, and certainly not, as Muenster would fain have us believe, that an Ostrich's dinner consists of a church-door key, and a horse-shoe. As matters of fact, we know that, when pursued, they do not bury their heads in the sand, or a bush; and instead of covering their eggs with sand, and leaving the sun to hatch them,

both the male and female are excellent, and model parents.

Pliny, however, says differently :—" This bird exceeds in height a man sitting on horseback, and can surpass him in swiftness, as wings have been given to aid it in running; in other respects Ostriches cannot be considered as birds, and do not raise themselves from the earth. They have cloven talons, very similar to the hoof of the stag (*they have but two toes*); with these they fight, and they also employ them in seizing stones for the purpose of throwing at those who pursue them. They have the marvellous property of being able to digest every substance without distinction, but their stupidity is no less remarkable : for although the rest of their body is so large, they imagine when they have thrust their head and neck into a bush, that the whole body is concealed."

Giovanni Leone Africano writes that " this fowle liveth in drie desarts and layeth to the number of ten or twelve egges in the sand, which being about the bignesse of great bullets weigh fifteen pounds a piece; but the ostrich is of so weak a memorie, that she presently forgetteth the place where her egges were laid, and, afterwards the same, or some other ostrich hen finding the said egges by chance hatched and fostereth them as if they were certainely her owne. The chickens are no sooner crept out of the shell but they prowle up and downe the desarts for their food, and before theyr feathers be growne they are so swifte that a man shall hardly overtake them. The ostrich is a silly and deafe creature, feeding upon any thing which it findeth, be it as hard and indigestible as yron."

The Halcyon.

Of this bird, the Kingfisher, Aristotle thus discourses:
—"The halcyon is not much larger than a sparrow; its colour is blue and green, and somewhat purple; its whole body is composed of these colours as well as the wings and neck, nor is any part without every one of these colours. Its bill is somewhat yellow, long and slight; this is its external form. Its nest resembles the marine balls which are called halosachnæ (*probably a Zoophyte*, Alcyonia) except in colour, for they are red; in form it resembles those sicyæ (cucumbers) which have long necks; its size is that of a very large sponge, for some are greater, others less. They are covered up, and have a thick solid part, as well as the cavity; it is not easily cut with a sharp knife, but, when struck or broken with the hand, it divides readily like the halosachnæ. The mouth is narrow, as it were a small entrance, so that the sea water cannot enter, even if the Sea is rough: its cavity is like that of the Sponge. The material of which the nest is composed is disputed, but it appears to be principally composed of the spines of the *belone*, for the bird lives on fish."

Pliny says:—"It is a thing of very rare occurrence to see a halcyon, and then it is only about the time of the setting of the Vergiliæ, and the summer and winter solstices; when one is sometimes to be seen to hover about a ship, and then immediately disappear. They hatch their young at the time of the winter solstice, from which circumstance those days are known as the 'halcyon days;' during this period the sea is calm and navigable, the Sicilian sea in particular."

"Halcyon days" is used proverbially, but the Kingfisher had another very useful trait. If a dead Kingfisher were hung up by a cord, it would point its beak to the quarter whence the wind blew. Shakespeare mentions this property in *King Lear* (ii. 1):—

> "Turn their halcyon beaks
> With every gale and vary of their masters."

And Marlowe, in his *Jew of Malta* (i. 1):—

> "But now, how stands the wind?
> Into what corner peers my halcyon bill?"

The Pelican.

The fable of the Pelican "in her piety, vulning herself," as it is heraldically described—is so well known,

as hardly to be worth mentioning, even to contradict it. In the first place, the heraldic bird is as unlike the real one, as it is possible to be; but the legend seems to have had its origin in Egypt, where the vulture was credited with this extraordinary behaviour, and this bird is decidedly more in accordance with the heraldic ideal. Du Bartas, singing of "Charitable birds," praises equally the Stork and the Pelican :—

> "The *Stork*, still eyeing her deer *Thessalie*,
> The *Pelican* comforteth cheerfully :

Prayse-worthy Payer; which pure examples yield
Of faithfull Father, and Officious Childe :
Th' one quites (in time) her Parents love exceeding,
From whom shee had her birth and tender breeding;
Not onely brooding under her warm brest
Their age-chill'd bodies bed-rid in the nest;
Nor only bearing them upon her back
Through th' empty Aire, when their own wings they lack;
But also, sparing (This let Children note)
Her daintiest food from her own hungry throat,
To feed at home her feeble Parents, held
From forraging, with heavy Gyves of Eld.
The other, kindly, for her tender Brood
Tears her own bowells, trilleth-out her blood,
To heal her young, and in a wondrous sort,
Unto her Children doth her life transport :
For finding them by som fell Serpent slain,
She rends her brest, and doth upon them rain
Her vitall humour; whence recovering heat,
They by her death, another life do get."

THE TROCHILUS.

This bird, as described by Aristotle, and others, is of a peculiar turn of mind :—"When the Crocodile gapes, the trochilus flies into its mouth to cleanse its teeth; in this process the trochilus procures food, and the other perceives it, and does not injure it; when the Crocodile wishes the trochilus to leave, it moves its neck that it may not bite the bird."

Giovanni Leone—before quoted—says, respecting this bird :—"As we sayled further we saw great numbers of crocodiles upon the banks of the ilands in the midst of Nilus lye baking them in the sunne with their jawes wide open, whereinto certaine little birds about the bignesse of a thrush entering, came flying forth againe presently after. The occasion whereof was told me to be this: the crocodiles by reason of their continuall devouring beasts and fishes have certaine pieces of flesh sticking fast betweene their forked teeth, which flesh being putrified, breedeth a kind of worme, wherewith they are cruelly tormented; wherefor the said birds flying about, and seeing the wormes enter into the Crocodile's jaws to satisfie their hunger thereon, but the Crocodile perceiving himselfe freede from the wormes of his teeth, offereth to shut his mouth, and to devour the little bird that did him so good a turne, but being hindred from his ungratefull attempt by a pricke which groweth upon the bird's head, hee is constrayned to open his jawes, and to let her depart."

Du Bartas gives another colour to the behaviour of the Trochilus :—

> "The *Wren*, who seeing (prest with sleep's desire)
> *Nile's* poys'ny Pirate press the slimy shoar,
> Suddenly coms, and, hopping him before,
> Into his mouth he skips, his teeth he pickles,
> Clenseth his palate, and his throat so tickles,
> That, charm'd with pleasure, the dull *Serpent* gapes.
> Wider and wider, with his ugly chaps :
> Then, like a shaft, th' *Ichneumon* instantly
> Into the Tyrants greedy gorge doth fly,
> And feeds upon that Glutton, for whose Riot,
> All *Nile's* fat margents scarce could furnish diet."

WOOLLY HENS.

Sir John Maundeville saw in "the kingdome named Mancy, which is the best kingdome of the worlde—(Manzi, *that part of China south of the river Hoang-ho*) whyte hennes, and they beare no feathers, but woll as shepe doe in our lande."

Two-Headed Wild Geese.

Near the land of the *Cynocephali* or dog-headed men, there were many islands, and, "Also in this yle, and in many yles thereabout are many wyld geese with two heads." But these were not the only extraordinary breed of wild geese, extant.

> "As the wise Wilde-geese, when they over-soar
> Cicilian mounts, within their bills do bear,
> A pebble stone both day and night : for fear
> Lest ravenous Eagles of the North descry
> Their Armies passage, by their Cackling Cry."

Aristotle mentions the Crane as another stone-bearing bird:—"Among birds, as it was previously remarked, the Crane migrates from one extremity of the earth to the other, and they fly against the wind. As for the story of the stone, it is a fiction, for they say that they carry a stone as ballast, which is useful as a touchstone for gold, after they have vomited it up."

Four-Footed Duck.

Gesner describes a four-footed duck, which he says is like the English puffin, except in the number of its feet: but Aldrovandus "out-Herods Herod" when he gives us "A monstrous Cock with Serpent's tail."

If we can believe Pliny, there are places where certain birds are never found:—"With reference to the departure of birds, the owlet, too, is said to lie concealed for a few days. No birds of this last kind are to be found in the island of Crete, and if any are imported thither, they immediately die. Indeed, this is a remarkable distinction made by Nature; for she denies to certain places, as it were, certain kinds of fruits and shrubs, and of animals as well; . . .

"Rhodes possesses no Eagles. In Italy, beyond the Padus, there is, near the Alps, a lake known by the name of Larius, beautifully situate amid a country covered with shrubs; and yet this lake is never visited by storks, nor, indeed, are they ever known to come within eight miles of it; whilst on the other hand, in the neighbouring territory of the Montres, there are immense flocks of magpies and jackdaws, the only bird that is guilty of stealing gold and silver, a very singular propensity.

"It is said that in the territory of Tarentum, the woodpecker of Mars is never found. It is only lately, too, and that but very rarely, that various kinds of pies have begun to be seen in the districts that lie between the Apennines, and the City; birds which are known by the name of *Variæ*, and are remarkable for the length of the tail. It is a peculiarity of this bird, that it becomes bald every year at the time of sowing rape. The partridge does not fly beyond the frontiers of Bœotia, into Attica; nor does any bird, in the island in the Euxine in which Achilles was buried, enter the temple there consecrated to him.

"In the territory of Fidenæ, in the vicinity of the City,

the storks have no young, nor do they build nests; but vast numbers of ring-doves arrive from beyond sea every year in the district of Volaterræ. At Rome, neither flies, nor dogs ever enter the temple of Hercules in the Cattle Market." . . .

Fish.

Terrestrial and Aerial animals were far more familiar to the Ancients than were the inhabitants of the vast Ocean, and not knowing much about them, their habits and ways, took "omne ignotum pro magnifico."

We have seen the union of Man and Beast, and Man and Bird; and Man and Fish was just as common, and perhaps more ancient than either of the former—for Berosus, the Chaldean historian, gives us an account of Oannes, or Hea, who corresponded to the Greek Cronos, who is identified with the fish-headed god so often represented on the sculptures from Nimroud, and of whom, clay figures have been found at Nimroud and Khorsabad, as well as numerous representations on seals and gems.

Of this mysterious union of Man and Fish, Berosus says:—" In the beginning there were in Babylon a great number of men of various races, who had colonised Chaldea. They lived without laws, after the manner of animals. But in the first year there appeared coming out of the Erythrian Sea (*Persian Gulf*) on the coast where it borders Babylonia, an animal endowed with reason, named Oannes. He had all the body of a fish, but below the head of the fish another head, which was that of a man; also the feet of a man, which came out of its fish's tail. He had a human voice, and its image is preserved to this day. This animal passed the day time among men, taking no nourishment. It taught them the use of letters, of sciences, and of arts of every kind; the rules for the foundation of towns, and the building of temples, the principles of laws, and geometry, the sowing of seeds, and the harvest; in one word, it gave to men all that conduced to the enjoyment of life. Since that time nothing excellent has been invented. At the time of sunset, this monster Oannes threw itself into the sea, and passed the night beneath the waves, for it was amphibious. He wrote a book upon the beginning of all things, and of Civilisation, which he left to mankind."

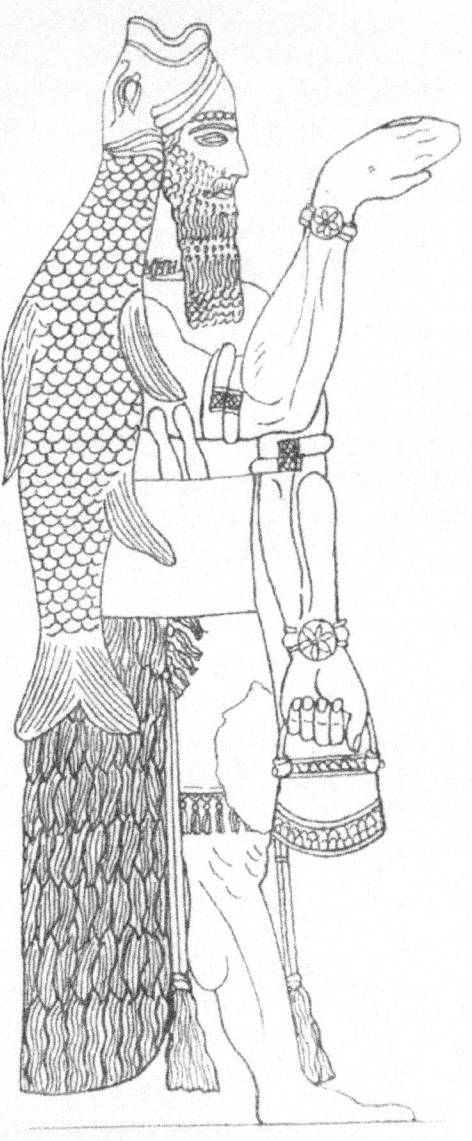

Helladice quotes the same story, and calls the composite being Oes; while another writer, Hyginus, calls him Euahanes. M. Lenormant thinks that it is evident that this latter name is more correct than Oannes, for it points to one of the Akkadian names of Hea—"Hea-Khan," *Hea, the fish*—and must be identified with the fish-God in the illustration.

Alexander Polyhistor, who mainly copied from Berosus, says that Oannes wrote concerning the generation of Mankind, of their different ways of life, and of their civil polity; and the following is the purport of what he wrote:—

"There was a time in which there existed nothing but darkness, and an abyss of waters, wherein resided most hideous beings, which were produced on a twofold principle. There appeared men, some of whom were furnished with two wings, others with four, and two faces. They had one body, but two heads; the one that of a man, the other of a woman; they were likewise in their several organs both male and female. Other human beings were to be seen with the legs and horns of a goat; some had horse's feet, while others united the hind-quarters of a horse with the body of a man, resembling in shape the hippocentaurs. Bulls likewise were bred then with the heads of men, and dogs with fourfold bodies, terminated in their extremities with the tails of fishes; horses also with the heads of dogs; men, too, and other animals, with the heads and bodies of horses, and the tails of fishes. In short, there were creatures in which were combined the limbs of every species of animals. In addition to these, fishes, reptiles, serpents, with other monstrous animals, which assumed each other's shape and countenance. Of all which were preserved delineations in the temple of Belus, at Babylon."

But, undoubtedly, the earliest representation of the *real* Merman—half-man, half-fish—comes to us from the uncovered palace of Khorsabad. On a portion of its sculptured walls is a representation of Sargon, the father of Sennacherib, sailing on his expedition to Cyprus, B.C. 720—on which occasion he had wooden images of

the gods made and thrown overboard in order to accompany him on his voyage. Among these is Hea, or Oannes, which I venture to assert is the first representation of a Merman.

In Hindoo Mythology, one of the incarnations, or *avatars* of Vishnu, represents him as issuing from the mouth of a fish. The God Dagon (Dag in Hebrew, signifying fish) was probably Oannes or Hea—and Atergatis was depicted as a Mermaid, half-woman, half-fish.

The Greeks worshipped her as Astarte, and later on as Venus Aphrodite she was perfect woman, still, however, born of the Sea-foam, and attended by Tritons or Mermen.

These Tritons and Nereids, male and female, were firmly believed in by both Greek and Roman—who both depicted them alike—the Triton, sometimes having a trident, sometimes without, but both Triton, and Nereid, perfect man and woman, of high types of manly and feminine beauty, to the waist—below which was the body of a fish of the Classical dolphin type. So ingrained have these forms become in humanity, that it would seem almost impossible to realise a Merman, or Mermaid, other than as usually depicted.

Pliny, of course, tells about them:—"A deputation of persons from Olisipo (*Lisbon*) that had been sent for the purpose, brought word to the Emperor Tiberius that a Triton had been both seen and heard in a certain cavern, blowing a Conch shell, and of the form they are usually represented. Nor yet is the figure generally attributed to the nereids at all a fiction, only in them the portion of the body that resembles the human figure, is still rough all over with scales. For one of these creatures was seen upon the same shores, and, as it died, its plaintive murmurs were heard, even by the inhabitants, at a distance.

"The legatus of Gaul, too, wrote word to the late Emperor Augustus, that a considerable number of nereids had been found dead upon the sea-shore. I have, too, some distinguished informants of equestrian rank, who state that they themselves once saw, in the Ocean of Gades, a sea-man, which bore in every part of his body, a perfect resemblance to a human being, and that during

the night he would climb up into ships; upon which the side of the vessel, where he seated himself, would instantly sink downward, and, if he remained there any considerable time, even go under water."

Ælian tells us, that it is reported that the great sea which surrounds the Island of Taprobana (*Ceylon*) contains an immense multitude of fishes and whales, and some of them have the heads of lions, panthers, rams, and other animals; and (which is more wonderful still) some of the Cetaceans have the form of Satyrs.

Gesner obligingly depicts this Pan, Sea Satyr, Ichthyo

centaurus, or Sea Demon, as he is indifferently called, and wants to pass it off as a veritable Merman, probably on account of its human-like trunk. He also quotes Ælian as to the authenticity of this monster,—and he gives a picture of another Man-fish, which he says was seen at Rome, on the third of November, 1523. Its size was that of a boy about five years of age. (See next page.)

Mermen and Mermaids do not seem to affect any particular district, they were met with all over the world— and records of their having been seen, come to us from all parts. That was well, and occurred in the ages of

CURIOUS CREATURES.

faith, but now the materialism of the present age would shatter, if it could, our cherished belief in these Marine eccentricities, and would fain have us to credit that all those that have been seen, were some of the Phocidæ, such as a "Dugong," or else they would attempt to persuade us that a beautiful mermaid, with her comb and looking-glass, was neither more nor less than a repulsive-looking "Manatee."

Sir J. Emerson Tennent quotes in his "Natural

History of Ceylon" from the description of one of the Dutch Colonial Chaplains, named Valentyn, who wrote an account of the Natural History of Amboyna. He says that in 1663, a lieutenant in the Dutch army was with some soldiers on the sea-beach at Amboyna, when they all saw mermen swimming near the beach. He described them as having long and flowing hair, of a colour between grey and green. And he saw them again, after an interval of six weeks, when he was in company with some fifty others. He also says that these Marine

Curiosities, both male and female, have been taken at Amboyna : and he cites a special one, of which he gives a portrait, that was captured by a district visitor of the Church, and presented by him to the Governor.

This last animal enjoyed European fame, as in 1716, whilst Peter the Great was the guest of the British Ambassador at Amsterdam, the latter wrote to Valentyn, asking that the marvel should be sent over for the Czar's inspection—but it came not. Valentyn also tells how, in the year 1404, a mermaid, tempest-tossed, was driven through a breach in a dyke at Edam, in Holland, and was afterwards taken alive in the lake of Parmen, whence she was carried to Haarlem. The good Dutch vrows took kindly care of her, and, with their usual thriftiness, taught her a useful occupation, that of spinning; nay, they Christianised her—and she died a Roman Catholic, several years after her capture.

The authentic records, if trust can be placed in them, are various and many—but are hardly worth recapitulating because of their sameness, and the smile of incredulity which their recital provokes.

Let us therefore turn to the monarch of the deep, the Whale—and of this creature we get curious glimpses from the Northern Naturalists ; but, before investigating this authentic denizen of ocean, we will examine some whose title to existence is not quite so clearly made out. Olaus Magnus gives us an introduction to some of "The horrible Monsters of the Coast of Norway. There are monstrous fish on the Coasts or Sea of *Norway*, of unusual Names, though they are reputed a kind of *Whales;* and, if men look long on them they will fright and amaze them. Their forms are horrible, their heads square, all set with prickles, and they have sharp and long Horns round about, like a tree rooted up by the roots: they are ten or twelve Cubits long, very black, and with huge eyes, the Compass whereof (i.e., *of the fish*) is above eight or ten Cubits: the apple of the eye is of one Cubit, and is red and fiery coloured, which in the dark night appears to Fisher-men afar off under Waters,

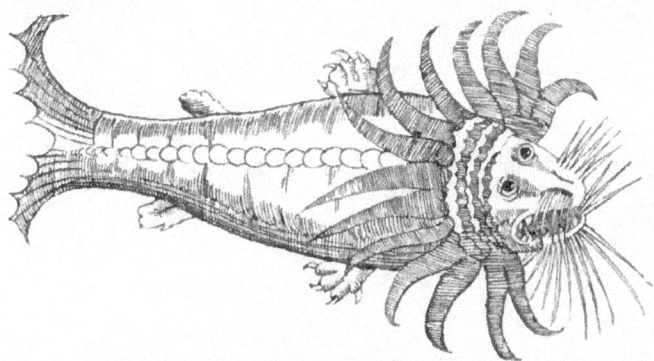

as a burning Fire, having hairs like Goose-Feathers, thick and long, like a beard hanging down; the rest of the body, for the greatness of the head, which is square, is very small, not being above fourteen or fifteen cubits long; one of these Sea Monsters will drown easily many great ships, provided with many strong Marriners."

He also speaks of a Cetacean, called a Physeter:— " The Whirlpool, or Prister, is of the kind of Whales, two hundred Cubits long, and is very cruel. For, to the danger of Sea men, he will sometimes raise himself beyond the Sail yards, and cast such floods of Waters above his head, which he had sucked in, that with a cloud of them, he will often sink the strongest ships, or expose the Marriners to extream danger. This Beast hath also a long and large round mouth like a Lamprey, whereby he sucks in his meat or water, and by his weight cast upon the Fore or Hinder-Deck, he sinks, and drowns a ship.

"Sometimes, not content to do hurt by water onely, as I said, he will cruelly over throw the ship like any small Vessel, striking it with his back, or tail. He hath a thick black Skin, all his body over; long fins, like to broad feet, and a forked tail 15 or 20 foot broad, wherewith he forcibly binds any parts of the ship, he twists it about. A Trumpet of War is the fit remedy

against him, by reason of the sharp noise, which he cannot endure: and by casting out huge great Vessels, that hinders this Monster's passage, or for him to play withall; or with Strong Canon and Guns, with the sound thereof he is more frighted, than with a Stone, or Iron Bullett; because this Ball loseth its force, being hindered by his Fat, or by the Water, or wounds but a little, his most vast body, that hath a Rampart of mighty Fat to defend it. Also, I must add, that on the Coasts of *Norway*, most frequently both Old and New Monsters are seen, chiefly by reason of the inscrutable depth of the Waters. Moreover, in the deep Sea, there are many kinds of fishes that are seldome or never seen by Man."

We have the saying, "Throw a tub to the Whale," and we not only find that it is the proper treatment to con-

ciliate Physeters, but Gesner shows us the real thing applied to Whales, trumpet and all complete, and he also shows us the close affinity between the Whale and the

Physeter, in the accompanying illustration, which depicts a whale uprearing, and coming down again on an unfortunate vessel.

There is another Whale, described by Gesner, which he calls the "Trol" whale, or in German, "Teüfelwal," or Devil Whale. This whale lies asleep on the water, and is of such a deceptive appearance that seamen mistake it for an island, and cast anchor into it, a proceeding which this peculiar class of whale does not appear to take much heed of. But, when it comes to lighting a fire upon it, and cooking thereon, it naturally

wakes up the whale. It is of this "Teüfelwal" that Milton writes ("Paradise Lost," Bk. i., l. 200):—

> " Or that sea-beast
> Leviathan, which God of all His works
> Created hugest that swim the ocean-stream.
> Him, haply slumbering on the Norway foam,
> The pilot of some small night-foundered skiff,
> Deeming some island, oft, as seamen tell,
> With fixèd anchor in his scaly rind,
> Moors by his side under the lee, while night
> Invests the sea, and wishèd morn delays."

And the same story is told in the First Voyage of Sindbad the Sailor, or, as Mr. Lane, whose translation (ed. 1883) I use, calls him, Es-Sindibád of the Sea:— "We continued our voyage until we arrived at an island like one of the gardens of Paradise, and at that island, the master of the ship brought her to anchor with us. He cast the anchor, and put forth the landing plank, and all who were in the ship landed upon that island. They had prepared for themselves fire-pots, and they lighted the fires in them, and their occupations were various: some cooked, others washed, and others amused themselves. I was among those who were amusing themselves upon the shores of the island, and the passengers were assembled to eat and drink, and play and sport. But while we were thus engaged, lo, the master of the ship, standing upon its side, called out with his loudest voice, 'O ye passengers, whom may God preserve! come up quickly into the ship, hasten to embark, and leave your merchandise, and flee with your lives, and save yourselves from destruction; for this apparent island upon which ye are, is not, in reality, an island, but it is a great fish that hath become stationary in the midst of the sea, and the sand hath accumulated upon it, so that it hath become like an island, and trees have grown upon it, since times of old; and, when ye lighted upon it the fire, it felt the heat, and put itself in motion, and now it will descend with you into the sea, and ye will all be drowned; then seek for yourselves escape before destruction, and leave the merchandise!' The passengers, therefore, hearing the words of the master of the ship, hastened to go up into the vessel, leaving the merchandise, and their other goods, and their copper cooking-pots, and their fire-pots; and some reached the ship, and others reached it not. The island had moved, and descended to the bottom of the sea, with all that were upon it, and the roaring sea, agitated with waves, closed over it."

Olaus Magnus, too, tells of sleeping whales being mistaken for islands:—"The Whale hath upon its Skin a superficies, like the gravel that is by the sea side; so that oft times when he raiseth his back above the waters, Sailors take it to be nothing else but an Island, and sayl unto it, and go down upon it, and they strike in piles upon it, and fasten them to their ships: they kindle fires to boyl their meat; until at length the Whale feeling the fire, dives down to the bottome; and such as are upon his back, unless they can save themselves by ropes thrown forth of the ship, are drown'd. This Whale, as I have said before of the Whirlpool and Pristes, sometimes so belcheth out the waves that he hath taken in, that, with a Cloud of Waters, oft times, he will drown the ship; and when a Tempest ariseth at Sea, he will rise above water, that he will sink the ships, during these Commotions and Tempests. Sometimes he brings up Sand on his back, upon which, when a Tempest comes, the Marriners are glad that they have found Land, cast Anchor, and are secure on a false ground; and when as they kindle their fires, the Whale, so soon as he perceives it, he sinks down suddenly into the depth, and draws both men and ships after him, unless the Anchors break."

But *apropos* of the whale casting forth such quantities of water, it is, as a matter of fact, untrue. The whale has a tremendously strong exhalation, and when it breathes under water, its breath sends up two columns of *spray*, but, if its head is above water, it cannot spout.

One thing in favour of whales, is "The Wonderful affection of the whales towards their young. Whales, that have no Gills, breathe by Pipes, which is found but in few creatures. They carry their young ones, when they are weak and feeble; and if they be small, they take them in at their mouths. This they do also when a Tempest is coming; and after the Tempest, they Vomit them up. When for want of water their young are hindered, that they cannot follow their Dams, the Dams take water in their mouths, and cast it to them like a

CURIOUS CREATURES.   221

river, that she may so free them from the Land they are fast upon. Also she accompanies them long, when

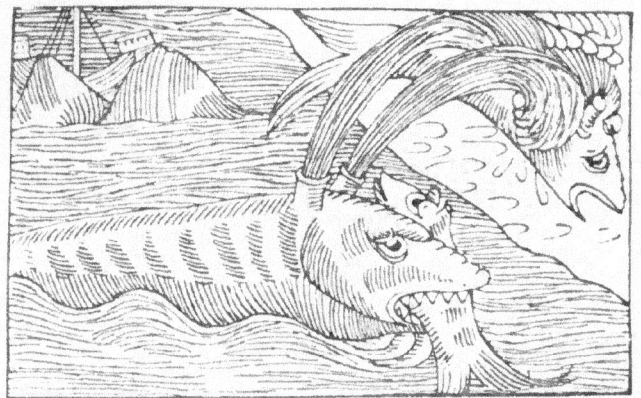

they are grown up; but they quickly grow up, and increase ten years."

According to Olaus Magnus, there be many kinds of whales :—" Some are hairy, and of four Acres in bigness; the Acre is 240 foot long and 120 broad; some are smooth skinned, and those are smaller, and are taken in the West and Northern Sea; some have their Jaws long and full of teeth; namely, 12 or 14 foot long, and the Teeth are 6, 8, or 12 foot long. But their two Dog teeth, or Tushes, are longer than the rest, underneath, like a Horn, like the teeth of Bores, or Elephants. This kind of whale hath a fit mouth to eat, and his eyes are so large, that fifteen men may sit in the room of each of them, and sometimes twenty, or more, as the beast is in quantity.

"His horns are 6 or 7 foot long, and he hath 250 upon each eye, as hard as horn, that he can stir stiff or gentle, either before or behind. These grow together, to defend his eyes in tempestuous weather, or when any other Beast that is his enemy sets upon him; nor is it a wonder, that he hath so many Horns, though they be

very troublesome to him; when, as between his eyes, the space of his forehead is 15 or 20 foot."

"The Spermaceti whale (*Physeter macrocephalus*) is the subject of a curious story, according to Olaus Magnus. He declares Ambergris is the sperm of the male Whale, which is not received by the female. "It is scattered wide on the sea, in divers figures, of a blew colour, but more tending to white; and these are glew'd together; and this is carefully collected by Marriners, as I observed, when, in my Navigation I saw it scattered here and there: This they sell to Physitians, to purge it; and when it is purged, they call it *Amber-greese*, and they use it against the Dropsie and Palsie, as a principal and most pretious unguent. It is white; and if it be found, that is of the colour of Gyp, it is the better. It is sophisticated with the powder of Lignum, Aloes, Styrax, Musk, and some other things. But this is discovered because that which is sophistocated will easily become soft as Wax, but pure *Amber-greese* will never melt so. It hath a corroborating force, and is good against swoundings and the Epilepsie."

As a matter of fact, it is believed to be a morbid secretion in the intestinal canal of the whale, originating in its bile. It is found in its bowels, and also floating on the sea, grey-coloured, in lumps weighing from half an ounce to one hundred pounds. Its price is about £3 per oz. It is much used in perfumery, but not in medicine, at least in Europe: but in Asia and Africa, it is, in some parts, so used, and also in cookery.

Olaus Magnus, too, tells us of the benefits the whale confers on the inhabitants of the cold and dreary North. How they salt the flesh for future eating, and the usefulness of the fat for lighting and warming through the long Arctic winter, while the small bones are used as fuel. Of the skin of this useful mammal, they make Belts, Bags, and Ropes, whilst a whole skin will clothe forty men. But these are not all its uses.

"Having spoken that the bodies of Whales are very large, for their head, teeth, eyes, mouth and skin; the bones require a place to be described; and it is thus. Because the vehemency of Cold in the farther parts of the North, and horrid Tempests there, will hardly suffer Trees to grow up tall, whereof necessary houses may be builded: therefore provident Nature hath provided for the Inhabitants, that they may build their houses of the most vast Ribs of Sea Creatures, and other things belonging thereunto. For these monsters of the Sea, being driven to land, either by some others that are their Enemies, or drawn forth by the frequent fishing for them by men, that the Inhabitants there may make their prey of them, or whether they die and consume; it is certain, that they leave such vast bones behind them, that whole Mansion Houses may be made of them, for Walls, Gates, Windows, Coverings, Seats, and for Tables also. For these Ribs are 20, 30, or more feet in length. Moreover the Back-bones, and Whirl-bones, and the Forked-bones of the vast head, are of no small bigness: and all these by the industry of Artists, are so fitted with Saws and Files, that the

Carpenter in Wood, joyn'd together with Iron, can make nothing more compleat.

"When, therefore, the flesh of this most huge Beast is eat and dissolved, onely his bones remain like a great Keel; and when these are purged by Rain, and the Ayr, they raise them up like a house, by the force of men that are called unto it. Then by the industry of the Master Builder, Windows being placed on the top of the house, or sides of the Whale, it is divided into many convenient Habitations; and gates are made of the same Beasts Skin, that is taken off long before, for that and some other use, and is hardened by the sharpness of the winds. Also a part within this Keel raised up like a house, they make several Hog Sties and places for other creatures, as the fashion is in other houses of Wood; leaving always under the top of this structure, a place for Cocks, that serve instead of Clocks, that men may be raised to their labour in the night, which is there continual in the Winter-time. They that sleep between these Ribs, see no other Dreams, than as if they were always toiling in the Sea-waves, or were in danger of Tempests, to suffer shipwreck."

Besides men, Whales had their foes, in the deep, and there was, according to Du Bartas, one very formidable and cunning enemy, in the shape of a bird :—

> " Meanwhile the *Langa*, skimming, (as it were,)
> The Ocean's surface, seeketh everywhere,
> The hugy Whale; where slipping in (by Art),
> In his vast mouth, shee feeds upon his Hart."

But it is cheering to find, on the authority of the same author, that he also has a helpful friend :—

> " As a great Carrak, cumbred and opprest
> With her-self's burthen, wends not East and West,
> Star-boord, and Lar-boord, with so quick Careers
> As a small Fregat, or swift Pinnass steers;
> And as a large and mighty limbed Steed,
> Either of *Friseland*, or of *German* breed,
> Can never manage half so readily,
> As *Spanish* Jennet, or light *Barbarie;*
> So the huge *Whale* hath not so nimble motion
> As smaller fishes that frequent the Ocean;
> But, sometimes, rudely 'gainst a Rock he brushes,
> Or in some roaring straight he blindly rushes,

And scarce could live a Twelve month to an end,
But for the little *Musculus* (his friend),
A little Fish, that, swimming still before,
Directs him safe from Rock, from shelf and shoar."

But we have only spoken of a very few varieties of Whales; some yet remain, which may be styled "fancy" Whales. At all events, they are lost to our times. Herodotus tells us that in the Borysthenes (*Dneiper*) were "large whales without any spinal bones, which they call Antacæi, fit for salting." Then, Gesner gives us varieties of Whales, of which we know nothing. There is the bearded and maned creature with a face somewhat resembling that of a human being, found only in the remotest North, and there is the hairy whale, *Cetum Capillatum vel Crinitum*, or *Germanice*, Haarwal, but no particulars of this curious creature are given.

He presents us with the image of a Cetacean, which he calls an Indian Serpent—but he evidently is so doubtful of the creature's authenticity that he tells us that Hieronimus Cardanus sent it formerly to him. He cannot quite make it out, with its monkey's head, and paws, but points out that it must be an aquatic animal, because of its tail.

In his *Addenda et Emendanda*, he gives, on the authority of Olaus Magnus, a picture of an unnamed

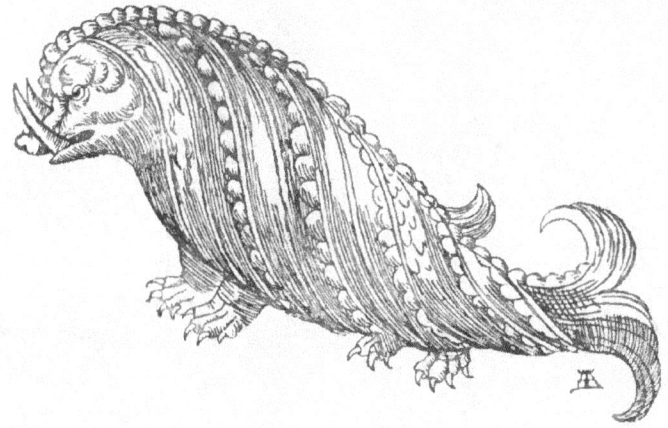

Whale—he says it was of great size, and had terrible teeth.

He also gives us two or three curious pictures of now extinct Cetaceans, something like terrestrial animals or men. And the first is a Leonine Monster, and for its authority he quotes Rondeletius.

This creature had none of its parts fitted to act as a marine animal of prey, but he says that Gisbertus (*Horstius*) Germanus, a physician at Rome, certifies that it was taken on the high seas, not long before the death of Pope Paul III., which took place A.D. 1549. It was of the size and shape of a Lion, it had four feet, not mutilated, or imperfect as those of the Seal, and not joined together as is the case with the beaver or duck,

but perfect, and divided into toes with nails: a long thin tail ending in hair; ears hardly visible, and its body covered with scales—but he adds that Gisbertus found fault with the artist, who had made the feet longer than they ought to have been—and the ears too large for an aquatic animal.

Gesner also gives us (and so does Aldrovandus) pictures of the Monk and Bishop fishes. The Monk-fish, he says, was caught off Norway, in a troubled sea: and he quotes Bœothius as describing a similar monster found in the Firth of Forth. The Bishop-fish was only *seen* off the coast of Poland, A.D. 1531.

The existence of these marine monsters had, at all

events, very wide credence, even if they never existed, for Sluper, whom I have before quoted, gives, in his

curious little book, two pictures of these two fishes (more awful than Gesner did). Of the Sea Monk he says:

> "La Mer poissons en abondance apporte,
> Par dons divins que devons estimer.
> Mais fort estrange est le Moyne de Mer,
> Qui est ainsi que ce pourtrait le porte."

And of the Sea Bishop:

> "La terre n'a Evesques seulement,
> Qui sŏt p bulle en grād hŏneur et tiltre,
> L'evesque croist en mer sembablement,
> Ne parlāt point, cōbien qu'il porte Mitre."

And Du Bartas writes of them, as if all in air, or on the earth, had its double in the sea—and he specially mentions these piscine ecclesiastics:—

CURIOUS CREATURES.   231

"Seas have (as well as skies) Sun, Moon, and Stars;
(As well as ayre) Swallows, and Rooks, and Stares;
(As well as earth) Vines, Roses, Nettles, Millions,[1]
Pinks, Gilliflowers, Mushrooms, and many millions
Of other Plants (more rare and strange than these)
As very fishes living in the Seas.
And also Rams, Calfs, Horses, Hares, and Hogs,
Wolves, Lions, Urchins, Elephants and Dogs,
Yea, Men and Mayds; and (which I more admire [2])
The mytred Bishop, and the cowled Fryer;
Whereof, examples, (but a few years since)
Were shew'n the Norways, and Polonian Prince."

Was the strange fish that Stow speaks of in his *Annales* one of these two?—"A.D. 1187. Neere unto Orforde in Suffolke, certaine Fishers of the sea tooke in their Nettes, a Fish having the shape of a man in all pointes, which Fish was kept by *Bartlemew de Glanville*, Custos of the castle of Orforde, in the same Castle, by the space of sixe monethes, and more, for a wonder: He spake not a word. All manner of meates he gladly did eate, but more greedilie raw fishe, after he had crusshed out all the moisture. Oftentimes he was brought to the Church where he showed no tokens of adoration. At length, when he was not well looked to, he stale away to the Sea and never after appeared." If this was not the real Simon Pure, yet I think it may put in a claim as a first-class British production, and, as far as I know, unique—all other denizens of the deep having some trace of their watery habitat, either in wearing scales, or a tail.

Following Du Bartas' idea, let us take some marine animals which have a somewhat similar counterpart on shore.

Gesner gives us the picture, Olaus Magnus gives us

[1] Melons.   [2] Wonder at.

the veracious history, of the Sea-cow:—"The Sea Cow is a huge Monster, strong, angry, and injurious; she brings forth a young one like to herself; yet not above two, but one often, which she loves very much, and leads it about carefully with her, whithersoever she swims to Sea, or goes on Land. Lastly this Creature is known to have lived 130 years, by cutting off her tail."

Olaus Magnus calls the Seal, the Sea-calf; and with trifling exceptions, gives a fair account of its habits, only there are some points which differ from the modern Seal, at all events:—"The Sea-Calf, which also in

Latine is called *Helcus*, hath its name from the likeness of a Land-Calf, and it hath a hard fleshy body; and therefore it is hard to be killed, but by breaking the Temples of the head. It hath a voice like a Bull, four feet, but not his ears; because the manner and mansion of its life is in the Waters. Had it such ears, they would take in much Water, and hinder the swimming of it. . . . They will low in their sleep, thence they are called Calves. They will learn, and with their voyce and countenance salute the company, with a confused murmuring; called by their names, they will answer, and no Creature sleeps more profoundly. The Fins that serve them for to swim in the Sea, serve for legs on Land, and they go hobling up and down as lame people do. Their Skins, though taken from their bodies, have always a sense of the Seas, and when the Sea goes forth, they will stand up like Bristles. The right Fin hath a soporiferous quality to make one sleep, if it be put under one's head. They that fear Thunder, think those Tabernacles best to live in, that are made of Sea-Calves Skins, because onely this Creature in the Sea, as an Eagle in the Ayr is safe and secure from the Stroke of Thunder. . . . If the Sea be boisterous and rise, so doth the Sea Calfe's hair: if the Sea be calm, the hair is smooth; and thus you may know the state of the Sea in a dead Skin. The *Bothnick* Marriners conjecture by their own Cloaths, that are made of these Skins, whether the Sea shall be calm, and their voyage prosperous, or they shall be in danger of Shipwreck. . . . These Creatures are so bold, that when they hear it thunder, and they see it clash and lighten, they are glad, and ascend upon the plain Mountains, as Frogs rejoyce against Rain."

A very fine piece of casuistry is shown, in "the perplexity of those that eat the flesh of *Sea-Calves* in *Lent*," and it seems to be finally settled that, according to "the men of a more clear judgment, rejecting many Reasons, brought on both sides, do say, and prove, that when the Sea-Calf brings forth on the shore, if the Beast driven by the Hunter, run into the Woods, men must forbear to eat of it in Lent, when flesh is forbidden; but if he run to the Waters, one may fairly eat thereof."

Gesner, in giving this delineation of a Sea-Horse, openly says that it is the Classical horse, as used by Neptunus; but Olaus Magnus declares that "The Sea Horse, between *Britany* and *Norway*, is oft seen to have a head like a horse, and to neigh; but his feet and hoof are cloven like to a Cow's; and he feeds both on Land, and in the Sea. He is seldome taken, though he

grow to be as big as an Ox. He hath a forked Tail like a Fish.

"THE SEA-MOUSE.

"The Sea-Mouse makes a hole in the Earth, and lays her Eggs there, and then covers them with Earth: on the 30th day she digs it open again, and brings her young to the Sea, first blind, and, afterwards, he comes to see.

"THE SEA HARE.

"The Sea-Hare is found to be of divers kinds in the Ocean, but so soon as he is caught, onely because he is suspected to be Venemous, how like so ever he is to a Hare, he is let loose again. He hath four Fins behind his Head, two whose motion is all the length of the fish, and they are long, like to a Hare's ears, and two again, whose motion is from the back, to the depth of the fishes belly, wherewith he raiseth up the weight of his head. This Hare is formidable in the Sea; on the

Land he is found to be as timorous and fearful as a hare."

The Sea-Pig.

Again we are indebted to Gesner for the drawing of this Sea Monster. Olaus Magnus, speaking of "The Monstrous Hog of the *German Ocean*," says:—"I spake before of a Monstrous Fish found on the Shores of *England*, with a clear description of his whole body, and every member thereof, which was seen there in the year 1532, and the Inhabitants made a Prey of it. Now I

shall revive the memory of that Monstrous Hog that was found afterwards, *Anno* 1537, in the same *German Ocean*, and it was a Monster in every part of it. For it had a Hog's head, and a quarter of a Circle, like the Moon, in the hinder part of its head, four feet like a Dragon's, two eyes on both sides in his Loyns, and a third in his belly, inclining towards his Navel; behind he had a forked Tail, like to other Fish commonly."

The Walrus.

Of the Walrus, Rosmarus, or Morse, Gesner draws, and Olaus Magnus writes, thus:—"The *Norway* Coast, toward the more Northern parts, hath a great Fish, as big as Elephants, which are called *Morsi*, or *Rosmari*, may be they are (called) so from their sharp biting; for, if they see any man on the Sea-shore, and can catch him, they come suddenly upon him, and rend him with their Teeth, that they will kill him in a trice. Therefore these Fish called *Rosmari*, or *Morsi*, have heads fashioned like to an Oxes, and a hairy Skin, and hair growing as thick as straw or corn-reeds, that lye loose very largely.

They will raise themselves with their Teeth, as by Ladders to the very tops of Rocks, that they may feed on the Dewie Grasse, or Fresh Water, and role themselves in it, unless in the mean time they fall very fast asleep, and rest upon the Rocks; for then Fishermen make all the haste they can, and begin at the Tail, and part the Skin from the Fat; and unto this that is parted, they put most strong Cords, and fasten them on the rugged rocks or Trees, that are near; then they throw stones at his head, out of a Sling, to raise him, and they compel him to descend, spoiled of the greatest part of his Skin, which is fastned to the Ropes: he being thereby debilitated, fearful, and half dead, he is made a rich prey, especially for his Teeth, that are very pretious amongst the *Scythians*, the *Muscovites*, *Russians*, and Tartars, (as Ivory amongst the Indians,) by reason of its hardness, whiteness, and ponderousnesse. For which Cause, by excellent industry of Artificers they are made fit for handles for Javelins: And this is also testified by *Mechovita*, an historian of *Poland*, in his double *Sarmatia*, and *Paulus Jovius* after him, relates it by the Relation of one *Demetrius*, that was sent from the great Duke of *Muscovy* to Pope Clement the 7th."

Although Olaus Magnus is very circumstantial in his detail as to the intense somnolence, and brutal flaying alive of the "thereby debilitated" Walrus, I can find no confirmation of either, in any other account—on the contrary, in "A Briefe Note of the Morse and the use thereof," published in Hakluyt, it is described as very wakeful and vigilant, and certainly not an animal likely to have salt put on its tail after Magnus's manner :—

"In the voyage of Jacques Carthier, wherein he discovered the Gulfe of S. Laurance, and the said Isle of Ramea in the yeere 1534, he met with these beastes, as he witnesseth in these words: About the said island are very great beasts as great as oxen, which have two great teeth in their mouthes like unto elephant's teeth, and live in the Sea. Wee sawe one of them sleeping upon the banks of the water, and, thinking to take it, we went to it with our boates, but so soon as he heard us, he cast himselfe into the sea. Touching these beasts which Jacques Carthier saith to be as big as oxen, and to have teeth in their mouthes like elephants teeth; true it is that they are called in Latine *Boves marini* or *Vaccæ marinæ*, and in the Russian tongue *morsses*, the hides whereof I have seene as big as any ox hide, and being dressed, I have yet a piece of one thicker than any two oxe, or bul's hides in England.

"The leather dressers take them to be excellent good to make light targets against the arrowes of the savages; and I hold them farre better than the light leather targets which the Moores use in Barbarie against arrowes and lances, whereof I have seene divers in her Majesties stately armourie in the Toure of London. The teeth of the sayd fishes, whereof I have seene a dry flat full at once, are a foote and sometimes more in length; and have been sold in England to the combe and knife makers at 8 groats and 3 shillings the pound weight, whereas the best ivory is solde for halfe the money; the graine of the bone is somewhat more yellow than the ivorie. One Mr. Alexander Woodson of Bristoll, my old friend, an excellent mathematician and skilful phisitian, shewed me one of these beasts teeth which were brought from the Isle of Ramea in the first prize, which was half a yard long, or very little lesse: and assured mee that he had made tryall of it in ministering medicine to his patients, and had found it as sovereigne against poyson as any unicorne's horne."

The Ziphius.

This Voracious Animal, whose size may be imagined by comparison with the Seal it is devouring, is thus described by Magnus:—"Because this Beast is conversant in the Northern Waters, it is deservedly to be joined with other monstrous Creatures. The Swordfish is like no other, but in something it is like a Whale. He hath as ugly a head as an Owl: his mouth is wondrous deep, as a vast pit, whereby he terrifies and drives away those that look into it. His

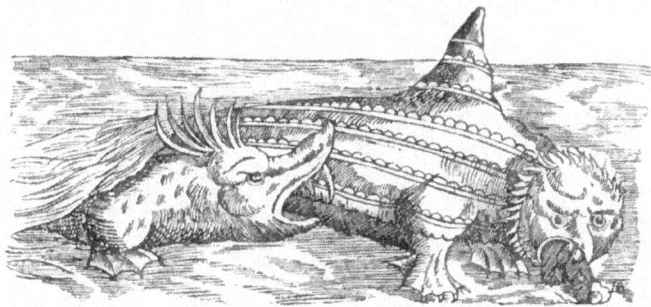

Eyes are horrible, his Back Wedge-fashion, or elevated like a Sword; his snout is pointed. These often enter upon the Northern Coasts as Thieves and hurtful Guests, that are always doing mischief to ships they meet, by boring holes in them, and sinking them.

"THE SAW FISH.

"The Saw fish is also a beast of the Sea; the body is huge great, the head hath a crest, and is hard and dented like to a Saw. It will swim under ships and cut them, that the Water may come in, and he may feed on the men when the ship is drowned."

THE ORCA

is probably the Thresher whale. Pliny thus describes it:—"The Balæna (*whale of some sort*) penetrates to our seas even. It is said that they are not to be seen in the ocean of Gades (*Bay of Cadiz*) before the winter solstice, and that at periodical seasons they retire and conceal themselves in some calm capacious bay, in which they take a delight in bringing forth. This fact, however, is known to the Orca, an animal which is particularly hostile to the Balæna, and the form of which cannot be in any way accurately described, but as an enormous mass of flesh, armed with teeth. This animal attacks the Balæna in its place of retirement, and with its teeth tears its young, or else attacks the females which have just brought forth, and, indeed, while they are still pregnant; and, as they rush upon them, it pierces them just as though they had been attacked by the beak of a Liburnian Galley. The female Balænæ, devoid of all flexibility, without energy to defend themselves, and overburdened by their own weight; weakened, too, by gestation, or else the pains of recent parturition, are well aware that their only resource is to take flight in the open sea, and to range over the whole face of the ocean; while the Orcæ, on the other hand, do all in their power to meet them in their flight, throw themselves in their way, and kill them either cooped up in a narrow passage, or else drive them on a shoal, or dash them to pieces against the rocks. When these battles are witnessed, it appears just as though the sea were infuriate against itself; not a breath of wind is there to be felt in the bay, and yet the waves, by their pantings and their repeated blows, are heaved aloft in a way which no whirlwind could effect.

"An Orca has been seen even in the port of Ostia, where it was attacked by the Emperor Claudius. It was while he was constructing the harbour there that this orca came, attracted by some hides, which, having been brought from Gaul, had happened to fall overboard there.

By feeding upon these for several days it had quite glutted itself, having made for itself a channel in the shoaly water. Here, however, the sand was thrown up by the action of the wind to such an extent that the creature found it quite impossible to turn round; and while in the act of pursuing its prey, it was propelled by the waves towards the shore, so that its back came to be perceived above the level of the water, very much resembling in appearance the keel of a vessel turned bottom upwards. Upon this, Cæsar ordered a number of nets to be extended at the mouth of the harbour, from shore to shore, while he himself went there with the Prætorian Cohorts, and so afforded a spectacle to the Roman people; for boats assailed the monster, while the soldiers on board showered lances upon it. I, myself, saw one of the boats sunk by the water which the animal, as it respired, showered down upon it."

Olaus Magnus thus writes "Of the fight between the Whale and the Orca. A *Whale* is a very great fish, about one hundred, or three hundred foot long, and the body is of a vast magnitude, yet the *Orca*, which is smaller in quantity, but more nimble to assault, and cruel to come on, is his deadly Enemy. An *Orca* is like a Hull turned inwards outward; a Beast with fierce Teeth, with which, as with the Stern of a Ship, he rends the *Whale's* Guts, and tears its Calve's body open, or he quickly runs and drives him up and down with his prickly back, that he makes him run to Fords and Shores. But the *Whale*, that cannot turn its huge body, not knowing how to resist the wily *Orca*, puts all its hopes in flight; yet that flight is weak, because this sluggish Beast, burdned by its own weight, wants one to guide her, to fly to the Foords, to escape the dangers."

The Dolphin.

Pliny says:—"The Dolphin is an animal not only friendly to man, but a lover of music as well; he is charmed by melodious concerts, and more especially by the notes of the water organ. He does not dread man, as though a stranger to him, but comes to meet ships, leaps and bounds to and fro, vies with them in swiftness, and passes them even when in full sail.

"In the reign of the late Emperor Augustus, a dolphin which had been carried to the Lucrine Lake, conceived a most wonderful affection for the child of a certain poor man, who was in the habit of going that way from Baiæ to Puteoli to school, and who used to stop there in the middle of the day, call him by his name of *Simo*, and would often entice him to the banks of the lake with pieces of bread which he carried for the purpose. At whatever hour of the day he might happen to be called by the boy, and although hidden and out of sight at the bottom of the water, he would instantly fly to the surface, and after feeding from his hand, would present his back for him to mount, taking care to conceal the spiny projection of his fins in their sheath, as it were; and so, sportively taking him up on his back, he would carry him over a wide expanse of sea to the school at Puteoli, and in a similar manner bring him back again. This happened for several years, until, at last, the boy happened to fall ill of some malady, and died. The Dolphin, however, still came to the same spot as usual, with a sorrowful air, and manifesting every sign of deep affliction, until at last, a thing of which no one felt the slightest doubt, he died purely of sorrow and regret.

"Within these few years also, another at Hippo Diarrhytus, on the coast of Africa, in a similar manner used to receive his food from the hands of various persons, present himself for their caresses, sport about among the swimmers, and carry them on his back. On being rubbed with unguents by Flavianus, the then pro-consul of Africa, he was lulled to sleep, as it appeared, by the sensation of an odour so new to him, and floated about just as though he had been dead. For some months after this, he carefully avoided all intercourse with man, just as if he had received some affront or other; but, at the end of that time, he returned, and afforded just the same wonderful scenes as before. At last, the vexations that were caused them by having to entertain so many influential men who came to see this sight, compelled the people of Hippo to put the animal to death. . . .

"Hegesidemus has also informed us, that, in the city of Iasus (*the island and city of Caria*), there was another boy also, Hermias by name, who in a similar manner used to traverse the sea on a dolphin's back, but that, on one occasion, a tempest suddenly arising, he lost his life, and was brought back dead: upon which, the dolphin, who thus admitted that he had been the cause of his death, would not return to the sea, but lay down upon dry land and there expired."

Du Bartas gives us a new trait in the Dolphin's character :—

> "Even as the Dolphins do themselves expose,
> For their live fellows, and beneath the waves
> Cover their dead ones under sandy graves."

The Narwhal,

generally called the Monoceros or Sea Unicorn, is thus shown in one place, by Gesner; and, rough though it is, it is far more like the Narwhal's horn than is the other, also, in his work, of a Sea Rhinoceros or

Narwhal engaged in combat with an outrageous - sized Lobster, or Kraken, I know not which; for, as we shall presently see, the Kraken is represented as a Crayfish or Lobster. It was the long twisted horn of the Narwhal which did duty for ages as the horn of the fabled Unicorn, a gift worthy to be presented by an Emperor to an Emperor.

This sketch of Gesner's, he describes as a one-horned monster with a sharp nose, devouring a Gambarus.

Olaus Magnus dismisses the Narwhal very curtly:—"The Unicorn is a Sea Beast, having in his forehead a very

great Horn, wherewith he can penetrate, and destroy the ships in his way, and drown multitudes of men. But divine goodnesse hath provided for the safety of Marriners herein; for, though he be a very fierce Creature, yet is he very slow, that such as fear his coming may fly from him."

The earlier voyagers who really saw the Narwhal, fairly accurately described it; as Baffin, whose name is so familiar to us by the bay called after him :—"As for the Sea Unicorne, it being a great fish, having a long horn or bone growing forth of his forehead or nostrill, such as Sir Martin Frobisher, in his second voyage found one, in divers places we saw them, which, if the horne be of any good value, no doubt but many of them may be killed;" and Frobisher, as reported in Hakluyt, says :—"On this west shore we found a dead fish floating, which had in his nose a horne streight, and torquet, (*twisted*) of length two yards lacking two ynches. Being broken in the top, here we might perceive it hollow, into the which some of our sailors, putting spiders, they presently died. I saw not the triall hereof, but it was reported unto me of a truth; by the vertue thereof we supposed it to be the Sea Unicorne."

THE SWAMFISCK.

The accompanying illustration, though heading the chapter in Olaus Magnus regarding the Swamfisck and other fish, does not at all seem to elucidate the text :—
" The Variety of these Fish, or rather Monsters, is here set down, because of their admirable form, and many properties of Nature, as they often come to the *Norway* Shores amongst other Creatures, and they are catcht

for their Fat, which they have in great plenty and abundance. For the Fisher-men purge it, by boyling it like flesh, on the fire, and they sell it to anoint leather, or for Oyl to burn in Lamps, to continue light, when it is perpetual darkness. Wherefore the first Monster that comes, is of a round form, in *Norway* called *Swam-fisck*, the greatest glutton of all other Sea-Monsters. For he is scarce satisfied, though he eat continually. He is said to have no distinct stomach; and so what he eats turns into the thickness of his body, that he

appears nothing else than one Lump of Conjoyned Fat. He dilates and extends himself beyond measure, and when he can be extended no more, he easily casts out fishes by his mouth because he wants a neck as other fishes do. His mouth and belly are continued one to the other. But this Creature is so thick, that when there is danger, he can, (like the Hedg-Hog) re-double his flesh, fat and skin, and contract and cover himself; nor doth he that but to his own loss, because fearing Beasts that are his Enemies, he will not open himself when he is oppressed with hunger, but lives by feeding on his own flesh, choosing rather to be consumed in part by himself, than to be totally devoured by Wild Beasts. If the danger be past, he will try to save himself.

"THE SAHAB.

"There is also another Sea-Monster, called *Sahab*, which hath small feet in respect of its great body, but he hath one long one, which he useth in place of a hand to defend all his parts; and with that he puts meat into his mouth, and digs up grass. His feet are almost gristly, and made like the feet of a Cow or Calf. This Creature swimming in the water, breathes, and when he sends forth his breath, it returns into the Ayr, and he casts Water aloft, as Dolphins and Whales do.

"THE CIRCHOS.

"There is also another Monster like to that, called *Circhos*, which hath a crusty and soft Skin, partly black, partly red, and hath two cloven places in his Foot, that serve for to make three Toes. The right foot of this Animal is very small, but the left is great and long; and, therefore, when he walks all his body leans on the left side, and he draws his right foot after him: When the Ayr is calm he walketh, but when the Wind is high, and the Sky cloudy, he applies himself to the Rocks, and rests unmoved, and sticks fast, that he can scarce be pulled off. The nature of this is wonderful enough: which in calm Weather is sound, and in stormy Weather is sick."

The Northern Naturalists did not enjoy the monopoly

of curious fish, for Zahn gives us a very graphic picture  of the different sides of two small fish captured in Denmark and Norway (*i.e.*, presumably in some northern region) with curious letters marked on them. He does not attempt to elucidate the writing; and as it is of no known language, we may charitably put it down to the original " Volapük." He also favours us with the effigies of a curious fish found in Silesia in 1609, also ornamented with an inscription in an unknown tongue.

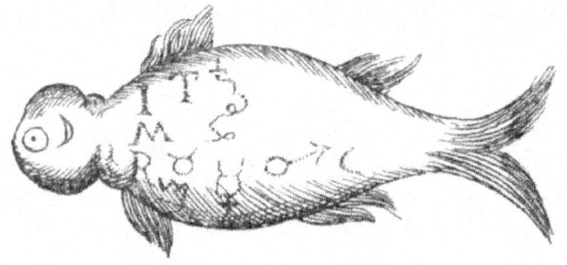

He also supplies us with the portrait of a pike, which was daintily marked with a cross on its side and a star on its forehead.

But too much space would be taken up if I were to recount all the piscine marvels that he relates.

Aristotle mentions that fish do not thrive in cold weather, and he says that those which have a stone in their head, as the chromis, labrax, sciæna, and phagrus, suffer most in the winter; for the refrigeration of the stone causes them to freeze, and be driven on shore.

CURIOUS CREATURES. 249

Sir John Mandeville, speaking of the kingdom of Talonach, says :—" And that land hath a marvayle that is

in no other land, for all maner of fyshes of the sea cometh there once a yeare, one after the other, and lyeth him neere the lande, sometime on the lande, and so lye three dayes, and men of that lande come thither and take of them what he will, and then goe these fyshes awaye, and another sort commeth, and lyeth also three dayes and men take of them, and do thus all maner of fyshes tyll all have bene there, and menne have taken what they wyll. And men wot not the cause why it is so. But they of that Countrey saye, that those fyshes come so thyther to do worship to theyr king, for they say he is the most worthiest king of the worlde, for he hath so many wives, and geateth so many children of them." (See next page.)

I know of no other fish of such an accommodating nature, except it be those of whom Ser Marco Polo speaks, when writing of Armenia :—" There is in this Country a certain Convent of Nuns called St. Leonard's, about which I have to tell you a very wonderful circumstance. Near the church in question there is a great lake at the foot of a mountain, and in this lake are found no fish, great or small, throughout the year till Lent come. On the first day of Lent they find in it the finest fish in the world, and great store, too, thereof;

and these continue to be found till Easter Eve. After that they are found no more till Lent come round again; and so 'tis every year. 'Tis really a passing great miracle!"

Edward Webbe, "Master Gunner," whose travels were printed in 1590, informs us that in the "Land of

Siria there is a River having great store of fish like unto Samon-trouts, but no Jew can catch them, though either Christian and Turk shall catch them in abundance, with great ease."

Pliny has some curious natural phenomena to tell us about, of showers of Milk, Blood, Flesh, Iron, and Wool; nay, he even says that, the year of this woolly shower, when Titus Annius Milo was pleading his own cause, there fell a shower of baked tiles!

After this we can swallow Olaus Magnus's story of a

rain of fishes very comfortably, especially as he supplements it with showers of frogs and worms.

He gives a curious story of the black river at the New Fort in Finland :—" There is a Fort in the utmost parts of *Finland* that is under the Pole, and it belongs to the Kingdom of *Sweden*, and it is called the New-Fort, because it was wonderfull cunningly built, and fortified by Nature and Art; for it is placed on a round Mountain, having but one entrance and outlet toward the West; and that by a ship that is tyed with great Iron Chains, which by strong labour and benefit of

Wheels, by reason of the force of the Waters, is drawn to one part of the River by night, by keepers appointed by the King of *Sweden*, or such as farm it. A vast river runs by this Castle, whose depth cannot be found; it ariseth from the White Lake, and falls down by degrees: at the bottome it is black, especially round this Castle, where it breeds and holds none but black Fish, but of no ill taste, as are Salmons, Trouts, Perch, Pikes, and other soft Fish. It produceth also the Fish *Trebius*, that is black in Summer, and white in Winter, who, as *Albertus* saith, grows lean in the Sea; but when he is a foot long, he is five fingers fat: This, seasoned with Salt, will draw Gold out of the deepest waters that it is fallen in, and make it flote from the bottom. At last, it makes the black Lake passing by *Viburgum*, as *Nilus* makes a black River, where he dischargeth himself.

"When the Image of a Harper, playing, as it were, upon his Harp, in the middle of the Waters above them

appears, it signifies some ill *Omen*, that the Governor of the Fort, or Captain shall suddenly be slain, or that the negligent and sleepy Watchman shall be thrown headlong from the high walls, and die by Martial Law. Also this water is never free from Ghosts and Visions that appear at all times; and a man may hear Pipes sound, and Cymbals tinkle, to the shore."

Aristotle mentions a fish called the Meryx that chewed the cud, and Pliny speaks of the Scarus, which, he says, "at the present day is the only fish that is said to ruminate, and feed on grass, and not on other fish." But he seems to have forgotten that in a previous place in the same book, he speaks of a large peninsula in the Red Sea, on the southern coast of Arabia, called Cadara, where "the sea monsters, just like so many cattle, were in the habit of coming on shore, and after feeding on the roots of shrubs, they would return; some of them, which had the heads of horses, asses, and bulls, found a pasture in the crops of grain."

The Remora.

Of this fish Pliny writes:—"There is a very small fish that is in the habit of living among the rocks, and is known as the Echeneis, Ἀπὸ τοῦ ἔχειν νῆας. (*From holding back ships.*) It is believed that when this has attached itself to the keel of a ship, its progress is impeded, and that it is from this circumstance that it takes its name. For this reason, also, it has a disgraceful repute, as being employed in love philtres, and for the purpose of retarding judgments and legal proceedings. . . . It is never used, however, for food. . . . Mucianus speaks of a Murex of larger size than the purple, with a head that is neither rough nor round; and the shell of which is single, and falls in folds on either side. He tells us, also, that some of these creatures once attached themselves to a ship freighted with children of noble birth, who were being sent by Periander for the purpose of being castrated, and that they stopped its course in full sail; and he further says, that the shell-fish which did this service are duly honoured in the temple of Venus, at Cnidos. Trebius Niger says that this fish is a foot in length, and five fingers in thickness, and that it can retard the course of vessels; besides which, it has another peculiar property—when preserved in salt, and applied, it is able to draw up gold which has fallen into a well, however deep it may happen to be."

> " But, *Clio*, wherefore art thou tedious
> In numbering *Neptune's* busie burgers thus?
> If in his works thou wilt admire the worth
> Of the Sea's Soverain, bring but only forth
> One little *Fish*, whose admirable story
> Sufficeth sole to shewe his might and glory.
> Let all the Windes, in one Winde gather them,
> And (seconded with *Neptune's* strongest stream)
> Let all at once blowe all the stiffest gales
> Astern a Galley under all her sails;
> Let her be holpen with a hundred Owers,
> Each lively handled by five lusty Rowers;
> The *Remora*, fixing her feeble horn
> Into the tempest beaten Vessel's Stern,
> Stayes her stone still, while all her stout Consorts
> Saile thence, at pleasure, to their wished Ports,
> Then loose they all the sheats, but to no boot:
> For the charm'd Vessell bougeth not a foot;
> No more than if, three fadom under ground,
> A score of Anchors held her fastly bound:
> No more than doth the Oak, that in the Wood,
> Hath thousand Tempests, (thousand times) withstood;
> Spreading as many massy roots belowe,
> As mighty arms above the ground do growe."

The Dog-fish and Ray.

Olaus Magnus writes of "The cruelty of some Fish, and the kindness of others. There is a fish of the kind of Sea-Dogfish, called *Boloma*, in *Italian*, and in *Norway*, *Haafisck*, that will set upon a man swimming in the Salt-Waters, so greedily, in Troops, unawares, that he will sink a man to the bottom, not only by his biting, but also by his weight; and he will eat his more tender parts, as his nostrils, fingers, &c., until such time as the Ray come to revenge these injuries; which runs thorow the Waters armed with her natural fins, and with some

violence drives away these fish that set upon the drown'd man, and doth what he can to urge him to swim out. And he also keeps the man, until such time as his spirit being quite gone; and after some days, as the Sea naturally purgeth itself, he is cast up. This miserable spectacle is seen on the Coasts of *Norway* when men go to wash themselves, namely, strangers and Marriners that are ignorant of the dangers, leap out of their ships into the sea. For these Dogfish, or *Boloma*, lie hid under the ships riding at Anchor as Water Rams, that they may catch men, their malicious natures stirring them to it."

THE SEA DRAGON.

Of the Ray tribe of fishes, the Sea Dragon is the most frightful-looking, but we know next to nothing about it. Pliny only cursorily mentions it thus :—" The Sea Dragon again, if caught, and thrown on the sand, works out a hole for itself with its muzzle, with the

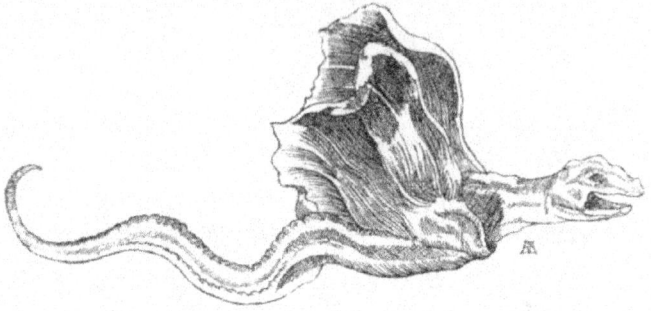

most wonderful celerity." Olaus Magnus simply copies Pliny almost word for word. Gesner, from whom I have taken this illustration, merely classes it among the Rays, and gives no further information about it; neither does Aldrovandus, from whom I have taken another picture.

THE STING RAY.

Pliny mentions the Sting Ray, and ascribes to it marvellous powers, which it does not possess :—" There is nothing more to be dreaded than the sting which protrudes from the tail of the *Trygon*, by our people known as the *Pastinaca*, a weapon five inches in length. Fixing this in the root of a tree, the fish is able to kill it; it can pierce armour, too, just as though with an arrow, and to the strength of iron it adds all the corrosive qualities of poison."

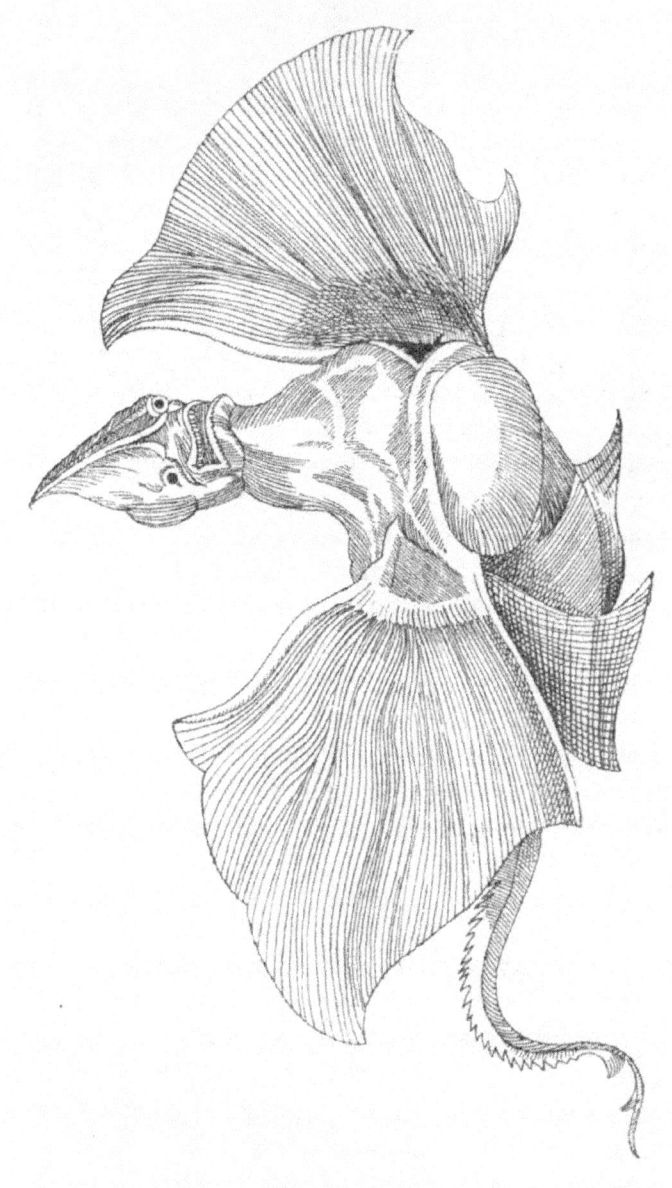

Senses of Fishes.

He also tells us about the senses of fishes, and first of their hearing:—"Among the marine animals, it is not probable that Oysters enjoy the sense of hearing, but it is said that immediately a noise is made, the Solen (*razor-sheath*) will sink to the bottom; it is for this reason, too, that silence is observed by persons while fishing at sea. Fishes have neither organs of hearing, nor yet the exterior orifice. And yet it is quite certain that they do hear, for it is a well-known fact, that in some fish-ponds they are in the habit of being assembled to be fed by the clapping of the hands. In the fish-ponds, too, that belong to the Emperor, the fish are in the habit of coming, each kind, as it bears its name. So, too, it is said the Mullet, the Wolf-fish, the Salpa, and the Chromis, have a very exquisite sense of hearing, and that it is for this reason that they frequent shallow water.

"It is quite manifest that fishes have the sense of smell also; for they are not all to be taken with the same bait, and are seen to smell at it before they seize it. Some, too, that are concealed in the bottom of holes are driven out by the fishermen, by the aid of the smell of salted fish; with this he rubs the entrance of their retreat in the rock, immediately upon which they take to flight from the spot, just as though they had recognized the dead carcases of those of their kind. Then, again, they will rise to the surface at the smell of certain odours, such, for instance, as roasted sepia and polypus; and hence it is that these baits are placed in the osier-kipes used for taking fish. They immediately take to flight upon smelling the bilge-water in a ship's hold, and more especially upon scenting the blood of fish.

"The Polypus cannot possibly be torn away from the rock to which it clings; but upon the herb *cunila* being applied, the instant it smells it, the fish quits its hold. . . . All animals have the sense of touch, those even which have no other sense; for even in the oyster, and, among land animals, in the worm, this sense is found. I am strongly inclined to believe, too, that the sense of taste exists in all animals; for why else should one seek one kind of food, and one another?"

ZOOPHYTES.

Writing on the lower phases of Marine Animal life, he says:—"Indeed, for my own part, I am strongly of opinion that there is sense existing in those bodies which have the nature of neither animals nor vegetables, but a third, which partakes of them both:—sea-nettles, and sponges, I mean. The Sea Nettle wanders to and fro by night, and at night changes its locality. These creatures are by nature a sort of fleshy branch, and are nurtured upon flesh. They have the power of producing an itching, smarting pain, just like that caused by the nettle found on land. For the purpose of seeking its prey, it contracts, and stiffens itself to the utmost possible extent, and then, as a small fish swims past, it will suddenly spread out its branches, and so seize and devour it. At another time it will assume the appearance of being quite withered away, and let itself be tossed to and fro, by the waves, like a piece of sea-weed, until it happens to touch a fish. The moment it does so, the fish goes to rub itself against a rock, to get rid of the itching: immediately upon which, the nettle pounces upon it. By night also it is on the look-out for Scallops and Sea-urchins. When it perceives a hand approaching it, it instantly changes its colour, and contracts itself; when touched, it produces a burning sensation, and if ever so short a time is afforded, makes its escape. Its mouth is situate, it is said, at the root or lower part, and the excrements are discharged by a small canal situated above.

"Sponges.

"We find three kinds of sponges mentioned; the first are thick, very hard, and rough, and are called *tragi:* the second are thick, and much softer, and are called *mani:* of the third, being fine, and of a closer texture, tents for sores are made; this last is known as *Achillium.* All of these sponges grow on rocks, and feed upon shell and other fish, and slime.

"It would appear that these creatures, too, have some intelligence; for, as soon as ever they feel the hand about to tear them off, they contract themselves, and are separated with much greater difficulty: they do the same also, when the waves buffet them to and fro. The small shells that are found in them, clearly show that they live upon food; about Torone it is even said that they will survive after they have been detached, and that they grow again from the roots which have been left adhering to the rock. They leave a colour similar to that of blood upon the rock from which they have been detached, and those, more especially, which are produced in the Syrtes of Africa."

Olaus Magnus gives us the accompanying illustration of Zoophytes and Sponges. Of the latter, he says:—
"Sponges are much multiplied near the Coasts of *Nor-*

CURIOUS CREATURES. 261

way; the nature of it is, that it agrees with other living creatures in the way of contracting, and dilating itself: yet some are immovable from rocks, and if they be broken off at the Roots, they grow again; some are

movable from place to place; and these are found in huge plenty on the foresaid shores. They are fed with mud, small fish, and oysters. When they are alive, they are black, as they are when they are wet."

The Kraken.

This enormous monster, peculiar to the Northern Seas, is scarcely a fable, because huge Calamaries are not infrequently seen. Poor Pontoppidan has often been considered a Danish Ananias, but there are authentic accounts of these enormous Cuttle-fish; for instance, in 1854, one was stranded at the Skag, in Jutland, which was cut in pieces by the fishermen in order to be used as bait, and filled many wheelbarrows. Another, either in 1860 or 1861, was stranded between Hillswick and Scalloway, on the west of Scotland, and its tentacles were sixteen feet long, the pedal arms about half as long, and its body seven feet. The French ship *Alecton*, on 30th November 1861, between Madeira and Teneriffe, slipped a rope with a running knot over an enormous calamary, but only brought a portion on board, the body breaking off. It was estimated at being sixteen to eighteen feet in length, without counting its arms. The legend of its sinking ships and taking sailors from them is common to many countries, even the Chinese and Japanese thus depicting them.

Olaus Magnus gives us a graphic picture of a huge Polyp, thus seizing a sailor, and dragging him from his ship in spite of all his efforts to prevent him. On next page is a huge calamary shown with a man in its clutches. This is both in Gesner and Aldrovandus. But this terror to mariners had its master in the Conger eel. Gesner, who has taken his picture from some description of the World, introduces it as a Sea-Serpent; but Aristotle says that "the Congers devour the Polypi, which cannot adhere to them on account of the

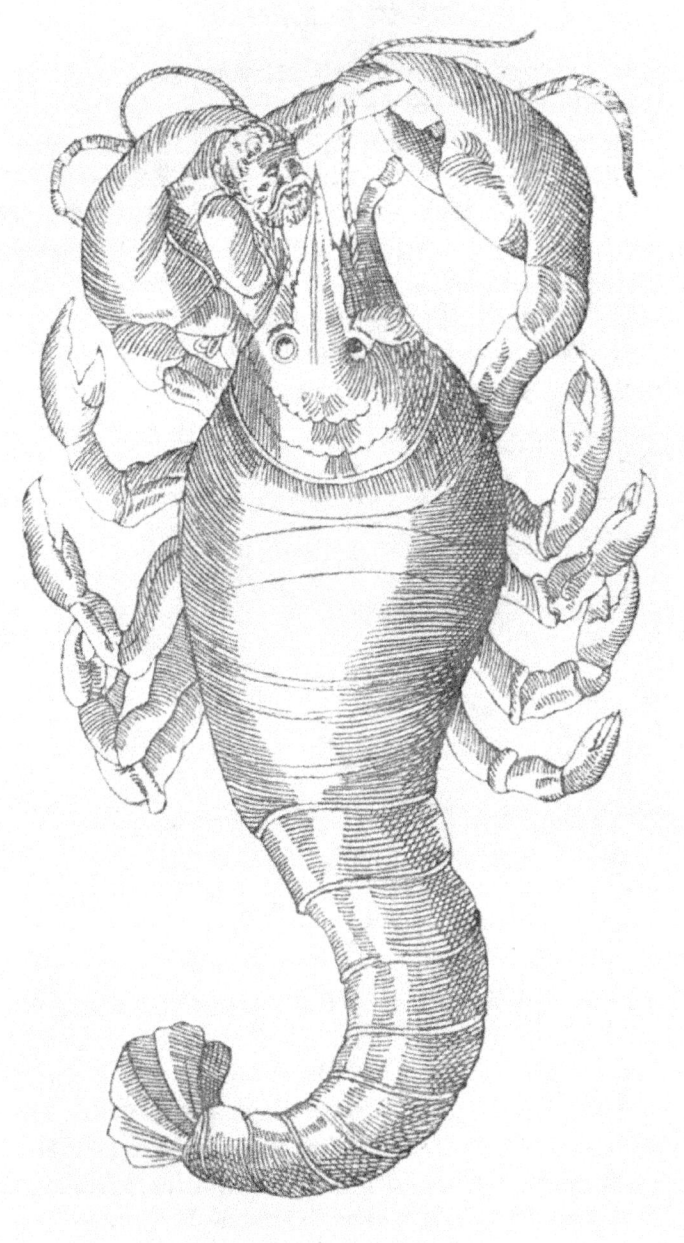

smoothness of their surface." Magnus also speaks of the antipathy between the two.

According to Pliny, quoting Trebius Niger, the Polypus shows a fair amount of cunning:—" Shell fish are destitute of sight, and, indeed, all other sensations but those which warn them of hunger, and the approach of danger. Hence it is that the Polypus lies in ambush till the fish opens its shell, immediately upon which, it places within it a small pebble, taking care, at the same time, to keep it from touching the body of the animal, lest,

by making some movement, it should chance to eject it. Having made itself thus secure, it attacks its prey, and draws out the flesh, while the other tries to contract itself, but all in vain, in consequence of the separation of the shell, thus effected by the insertion of the wedge.

"In addition to the above, the same author states that there is not an animal in existence, that is more dangerous for its powers of destroying a human being when in the water. Embracing his body, it counteracts his struggles, and draws him under with its feelers and its numerous suckers, when, as often is the case, it happens to make an attack upon a shipwrecked mariner or a child. If, however, the animal is turned over, it loses all its power; for when it is thrown upon its back, the arms open of themselves.

"The other particulars which the same author has given, appear still more closely to border upon the marvellous. At Carteia, in the preserves there, a Polypus was in the habit of coming from the sea to the pickling tubs that were left open, and devouring the fish laid in salt there—for it is quite astonishing how eagerly all sea animals follow even the very smell of salted condiments, so much so, that it is for this reason that the fishermen take care to rub the inside of the wicker fish-kipes with them.—At last, by its repeated thefts and immoderate depredations, it drew down upon itself the wrath of the keepers of the works. Palisades were placed before them, but these the Polypus managed to get over by the aid of a tree, and was only caught at last by calling in the assistance of trained dogs, which surrounded it at night, as it was returning with its prey; upon which, the keepers, awakened by the noise, were struck with alarm at the novelty of the sight presented.

"First of all, the size of the Polypus was enormous beyond all conception: and then it was covered all over with dried brine, and exhaled a most dreadful stench. Who could have expected to find a Polypus there, or could have recognised it as such, under these circumstances? They really thought that they were joining battle with some monster, for at one instant, it would drive off the dogs by its horrible fumes, and lash at them with the extremities of its feelers; while at another, it would strike them with its stronger arms, giving blows with so many clubs, as it were; and it was only with the greatest difficulty that it could be dispatched with the aid of a considerable number of three-pronged fish-spears. The head of this animal was shewn to Lucullus; it was in size as large as a cask of fifteen amphoræ (*about* 135 *gallons*), and had a beard (*iti tentaculæ*), to use the expression of Trebius himself, which could hardly be encircled with both arms, full of knots, like those upon a club, and thirty feet in length; the suckers, or calicules, as large as an urn, resembled a basin in shape, while the teeth again were of a corresponding largeness: its remains, which were carefully preserved as a curiosity, weighed seven hundred pounds."

Olaus Magnus says:—" On the Coasts of *Norway* there is a Polypus, or creature with many feet, which hath a pipe on his back, whereby he puts to Sea, and he moves that sometimes to the right, and sometimes to the left. Moreover, with his Legs as it were by hollow places, dispersed here and there, and by his Toothed Nippers, he fastneth on every living Creature that comes near to him, that wants blood. Whatever he eats he heaps up in the holes where he resides: Then he casts out the Skins, having eaten the flesh, and hunts after fishes that swim to them: Also he casts out the shels, and hard outsides of Crabs that remain. He changeth his colour by the colour of the stone he sticks unto, especially when he is frighted at the sight of his Enemy, the Conger. He hath 4 great middle feet, in all 8; a little body, which the great feet make amends for. He hath also some small feet that are shadowed and can scarce be perceived. By these he sustains, moves, and defends himself, and takes hold of what is from him: and he lies on his back upon the stones, that he can scarce be gotten off, onlesse you put some stinking smell to him."

Crayfish and Crabs.

Pliny tells us that in the Indian Ocean are Crayfish four cubits in length (six feet), and he claims for crabs a sovereign specific against bites of scorpions and snakes:—" River-Crabs taken fresh and beaten up and drunk in water, or the ashes of them, kept for the purpose, are useful in all cases of poisoning, as a counter poison; taken with asses' milk they are particularly serviceable as a neutralizer of the venom of the scorpion; goat's milk or any other kind of milk being substituted, where asses' milk cannot be procured. Wine, too, should also be used in all such cases. River-Crabs beaten up with Ocimum, and applied to Scorpions, are fatal to them. They are possessed of similar virtues, also, for the bites of all other kinds of venomous animals, the Scytale in particular, adders, the sea hare, and the bramble frog. The ashes of them, preserved, are good for persons who give symptoms of hydrophobia after being bitten by a mad dog, some adding gentian as well, and administering the mixture in wine. In cases, too, where hydrophobia has already appeared, it is recommended, that these ashes should be kneaded up into boluses with wine and swallowed. If ten of these crabs be tied together with a handful of Ocimum, all the scorpions in the neighbourhood, the magicians say, will be attracted to the spot. They recommend, also, that to wounds inflicted by the scorpion, these crabs, or the ashes of them, should be applied with Ocimum. For all these purposes, however, sea crabs, it should be remembered, are not so useful. Thrasyllus informs us that there is nothing so antagonistic to serpents as crabs: that swine, when stung by a serpent, cure themselves by eating them; and that, when the sun is in the sign of Cancer, serpents suffer the greatest tortures. . . .

"It is said that while the sun is passing through the sign of Cancer, the dead bodies of the crabs, which are lying on the shore, are transformed into serpents."

The Sea-Serpent.

Of the antiquity of the belief in the Sea-Serpent there can be no doubt, for it is represented on the walls of the Assyrian palace at Khorsabad, more than once, in the sculpture representing the voyage of Sargon to

Cyprus, thus giving it an authentic antiquity of over 2600 years: but as its existence must then have been a matter of belief, it naturally comes that it must be much older than that.

Aristotle, who wrote nearly 400 years later, speaks of them, and their savage disposition :—"In Libya, the serpents, as it has been already remarked, are very large. For some persons say that as they sailed along the coast, they saw the bones of many oxen, and that it was evident to them that they had been devoured by the serpents. And, as the ships passed on, the serpents attacked the triremes, and some of them threw themselves upon one of the triremes, and overturned it."

These, together with Sargon's Sea-Serpent, were doubtless marine snakes, which are still in existence, and are found in the Indian Ocean, but the larger ones seem to have been seen in more northern waters. It has been the fashion to pooh-pooh the existence of this sea monster, but there are many that still do believe in it most thoroughly; only, to express that belief would be to certainly expose oneself to ridicule. No one doubts the *bonâ fides* of those who narrate having seen them, but some one is sure to come forward with his pet theory as to its being a school of porpoises, or an enormous cuttle-fish, with its tentacles playing on the surface of the water; so that no one likes to confess that he has seen it.

Both Olaus Magnus and Gesner give illustrations of the Sea-Serpent of Norway, and I give that of the latter, as it is the best. The former says:—" They who

Work of Navigation, on the Coasts of *Norway*, employ themselves in fishing, or merchandize, do all agree in this strange Story, that there is a Serpent there which is of a Vast Magnitude, namely 200 feet long, and, moreover, 20 foot thick; and is wont to live in Rocks and Caves toward the Sea Coast about *Berge;* which will go alone from his holes in a clear night in Summer, and devour Calves, Lambs, and Hogs, or else he goes into the Sea to feed on Polypus, Locusts, and all sorts of Sea Crabs. He hath commonly hair hanging from his neck a cubit long, and sharp Scales, and is black, and he hath flaming shining eys. This Snake disquiets the Shippers, and he puts up his head on high like a pillar, and catcheth away men, and he devours them; and this hapneth not, but it signifies some wonderful change of the Kingdom near at hand; namely, that the Princes shall die, or be banished; or some Tumultuous Wars shall presently follow. There is also another Serpent of an incredible magnitude in a town called *Moos*, of the Diocess of *Hammer*; which, as a Comet portends a change in all the World, so, that portends a change in the Kingdom of *Norway*, as it was seen, *Anno* 1522, that lifts himself high above the Waters, and rouls himself round like a sphere. This Serpent was thought to be fifty Cubits long by conjecture, by sight afar off: there followed this the banishment of King *Christiernus*, and a great persecution of the Bishops; and it shew'd also the destruction of the Country."

Topsell, in his *Historie of Serpents*, 1608, does not add much to Sea-Serpent lore, but he adds the picture of another kind of Serpent, as does also Aldrovandus, whose illustration I give. (See p. 272.) Erik Pontoppidan, Bishop of Bergen, in his *Natürlichen Historie von Norwegen*, gives a picture of the Sea-Serpent, somewhat similar to that previously given by Hans Egede, "the Apostle of Greenland." (See next page.) Pontoppidan tried to sift the wheat from the chaff, in connection with the Natural History of the North, but he was not always successful. He gives several cases, one seemingly very well authenticated, of the appearance of Sea-Serpents.

But possibly more credence may be given to more modern instances. Sir Walter Scott, in the Notes to *The*

Pirate, says (speaking of Shetland and Orkney fishermen) :—" The Sea-Snake was also known, which, arising out of the depths of the ocean, stretches to the skies his enormous neck, covered with a mane like that of a war-horse, and with his broad glittering eyes, raised mast-head high, looks out, as it seems, for plunder or for victims." "The author knew a mariner, of some reputation in his class, vouch for having seen the celebrated Sea-Serpent. It appeared, as far as could

be guessed, to be about a hundred feet long, with the wild mane and fiery eyes which old writers ascribe to the monster; but it is not unlikely the spectator might, in the doubtful light, be deceived by a good Norway log on the water."

Mr. Maclean, the pastor of Eigg, an island in the Small Isles parish, Inverness-shire, wrote, in 1809, to Dr. Neill, the Secretary of the Wernerian Society, that he had seen a Sea-Serpent, while he was in a boat about two miles from land. The serpent followed the boat, and the minister escaped by getting on to a rock. He

described it as having a large head and slender tail, with no fins, its body tapering to its tail. It moved in

CURIOUS CREATURES.

undulations, and he thought its length might be seventy to eighty feet. It was seen, also, by the crews of thirteen fishing-boats, who, being frightened thereat, fled to the nearest creek for safety.

A Sea-Serpent, judged to be of the length of about eighty feet, was seen by a party of British officers, in Margaret's Bay, whilst crossing from Halifax to Mahone Bay, on 15th May 1833.

In 1847 a Sea-Serpent was seen frequently, in the neighbourhood of Christiansand and Molde, by many persons, and by one Lars Johnöen, fisherman at Smolen, especially. He said that one afternoon, in the dog-days, when sitting in his boat, he saw it twice in the course of two hours, and quite close to him. It came, indeed, to within six feet of him, and, becoming alarmed, he commended his soul to God, and lay down in the boat, only holding his head high enough to enable him to observe the monster. It passed him, disappeared, and returned; but a breeze springing up, it sank, and he saw it no more. He described it as being about six fathoms (thirty-six *feet*) long, the body (which was as round as a serpent's) two feet across, the head as long as a ten-gallon cask, the eyes round, red, sparkling, and about five inches in diameter; close behind the head, a mane, like a fin, commenced along the neck, and spread itself out on both sides, right and left, when swimming. The mane, as well as the head, was of the colour of mahogany. The body was quite smooth, its movements occasionally fast and slow. It was serpent-like, and moved up and down. The few undulations which those parts of the body and tail that were out of water made, were scarce a fathom in length. His account was confirmed by several people of position, a Surgeon, a

Rector, and a Curate, being among those who had seen a Sea-Serpent.

But an appearance of the Sea-Serpent, without doubt, is most satisfactorily attested by the captain and officers of H.M.S. *Dædalus*. The first notice of it was in the *Times* of 10th October 1848, in which was a paragraph, dated 7th October, from Plymouth :—

"When the *Dædalus* frigate, Captain M'Quhæ, which arrived here on the 4th inst., was on her passage home from the East Indies, between the Cape of Good Hope and St. Helena, her captain, and most of her officers and crew, at four o'clock one afternoon, saw a Sea-Serpent. The creature was twenty minutes in sight of the frigate, and passed under her quarter. Its head appeared about four feet out of the water, and there was about sixty feet of its body in a straight line on the surface. It is calculated that there must have been under water a length of thirty or forty feet more, by which it propelled itself at the rate of fifteen miles an hour. The diameter of the exposed part of the body was about sixteen inches; and when it extended its jaws, which were full of large jagged teeth, they seemed sufficiently capacious to admit of a tall man standing upright between them. The ship was sailing north at the rate of eight miles an hour. The *Dædalus* left the Cape of Good Hope on the 30th of July, and reached St. Helena on the 16th of August."

Captain M'Quhæ sent the following letter to Admiral Sir W. H. Gage, G.C.H., at Devonport :—

"HER MAJESTY'S SHIP *DÆDALUS*, HAMOAZE,
Oct. 11, 1848.

"SIR,—In reply to your letter of this day's date, requiring information as to the truth of a statement published in the *Times* newspaper, of a Sea-Serpent of extraordinary dimensions having been seen from Her Majesty's Ship *Dædalus*, under my command, on her passage from the East Indies, I have the honour to acquaint you, for the information of my Lords Commissioners of the Admiralty, that at five o'clock P.M., on the 6th of August last, in latitude 24° 44′ S. and longitude 9° 22′ E., the weather dark and cloudy, wind fresh from the N.W., with a long ocean swell from the S.W., the ship on the port tack heading N.E. by N., something very unusual was seen by Mr. Sartoris, midshipman, rapidly approaching the ship from before the beam. The circumstance was immediately reported by him to the officer of the watch, Lieutenant Edgar Drummond, with whom, and Mr. William Barrett, the master, I was at the time walking the quarter-deck. The ship's company were at supper.

"On our attention being called to the object, it was discovered to be an enormous Serpent, with head and shoulders kept about four feet constantly above the surface of the sea; and, as nearly as we could approximate by comparing it with the length of what our maintopsail-yard would show in the water, there was, at the very least, sixty feet of the animal *à fleur d'eau*, no portion of which was, to our perception, used in propelling it through the water, either by vertical or horizontal undulation. It passed rapidly, but so close under our lee quarter that, had it been a man of my acquaintance, I should have easily recognised his features with the naked eye; and it did not, either in approaching the ship or after it had passed our wake, deviate in the slightest degree from its course to the S.W., which it held on at the pace of from twelve to fifteen miles per hour, apparently on some determined purpose.

"The diameter of the Serpent was about fifteen or sixteen inches behind the head, which was, without any doubt, that of a snake; and it was never, during the twenty minutes that it continued in sight of our glasses, once below the surface of the water. Its colour, a dark brown, with yellowish white about the throat. It had no fins, but something like the mane of a horse, or rather a bunch of seaweed, washed about its back. It was seen by the quartermaster, the boatswain's mate, and the man at the wheel, in addition to myself and officers above mentioned.

"I am having a drawing of the Serpent made from a sketch taken immediately after it was seen, which I hope to have ready for transmission to my Lords Commissioners of the Admiralty by to-morrow's post.—I have, &c., PETER M'QUHÆ, CAPTAIN."

Space will not allow me to chronicle all the other appearances of Sea-Serpents from 1848 to the present time. Suffice it to say, they are not very uncommon, and as for veracity, I will give another instance of its being seen on board the Royal Yacht *Osborne*, on 2nd June 1877, off Cape Vito, Sicily. Lieutenant Haynes made sketches, and wrote a description, of it, which was confirmed by the Captain and several officers. He wrote :—

"Royal Yacht *Osborne*, Gibraltar,
June 6, 1877.

"On the evening of that day (June 2), the sea being perfectly smooth, my attention was first called by seeing a ridge of fins above the surface of the water extending about thirty feet, and varying from five to six feet in height. On inspecting it by means of a telescope, at about one and a half cable's distance, I distinctly saw a head, two flappers, and about thirty feet of an animal's shoulder.

"The head, as nearly as I could judge, was about six feet thick, the neck narrower, about four or five feet, the shoulder about fifteen feet across, and the flappers each about fifteen feet in length. The movements of the flappers were those of a turtle, and the animal resembled a huge seal, the resemblance being strongest about the back of the head. I could not see the length of the head, but from its crown or top to just below the shoulder (where it became immersed) I should reckon about fifty feet. The tail end I did not see, being under water, unless the ridge of fins to which my attention was first attracted, and which had disappeared by the time I got a telescope, were really the continuation of the shoulder to the end of the object's body. The animal's head was not always above water, but was thrown upwards, remaining above for a few seconds at a time, and then disappearing. There was an entire absence of 'blowing' or 'spouting.'"

I think the verdict may be given that its existence, although belonging to "Curious Zoology," is not impossible, and can hardly be branded as a falsehood.

Serpents.

Of Serpents Topsell has written a "Historie," which, if not altogether veracious, is very amusing; and I shall quote largely from it, as it shows us "the latest thing out" in Serpents as believed in, and taught, in the time of James I. He begins, of course, with their creation, and the Biblical mention of them, and then passes to the power of man over them in charming and taming them. Of the former he tells the following tale:—

"*Aloisius Cadamustus*, in his description of the New World, telleth an excellent hystorie of a *Lygurian* young Man, beeing among the *Negroes* travailing in *Affrick*, whereby he endeavoureth to proove, how ordinary and familiar it is to them, to take and charme Serpents.

"The young man beeing in *Affricke* among the *Negroes*, and lodged in the house of a Nephew to the Prince of *Budoniell*, when he was taking himselfe to his rest, suddenly awakened by hearing the unwonted noise of the hissing of innumerable sorts of Serpents; wherat he wondred, and beeing in some terror, he heard his Host (the Prince's Nephew) to make himselfe readie to go out of the doores, (for he had called up his servants to sadle up his Cammels:) the young man demaunded of him the cause, why he would go out of doores now so late in the darke night? to whom he answered, I am to goe a little way, but I will returne againe verie speedily; and so he went, and with a charme quieted the Serpents, and drove them all away, returning againe with greater speed than the *Lygurian* young man, his ghest, expected. And when he had returned, he asked his ghest if hee did not heare the inmoderate hyssing of the Serpents? and he answered, that he had heard them to his great terrour. Then the Prince's Nephew (who was called Bisboror) replyed, saying, they were Serpents which had beset the house, and would have destroyed all their Cattell and Heards, except hee had gone foorth to drive them away by a Charme, which was very common and ordinary in those parts, wherin were abundance of very hurtfull Serpents.

"The Lygurian young man, hearing him say so, marvailed above measure, and said, that this thing was so rare and miraculous, that scarcely Christians could beleeve it. The *Negro* thought it as strange that the young man should bee ignorant heereof, and therefore told him, that their Prince could worke more strange things by a Charme which he had, and that this, and such like, were small, vulgar, and not be counted miraculous. For, when he is to use any strong poyson upon present necessitie, to put any man to death, he putteth some venom uppon a sword, or other peece of Armour, and then making a large round Circle, by his Charme compelleth many Serpents to come within that circle, hee himselfe standing amongst them, and observing the most venomous of them all so assembled, which he thinketh to contain the strongest poyson, killeth him, and causeth the residue to depart away presentlie; then, out of the dead Serpent hee taketh the poyson, and mixeth it with the seede of a certaine vulgar Tree, and therewithall annoynteth his dart, arrow, or sword's point, whereby is caused present death, if it give the bodie of a man but a very small wound, even to the breaking of the skinne, or drawing of the blood. And the saide *Negro* did earnestly perswade the young man to see an experiment hereof, promising him to shew all as he had related, but the *Lygurian* beeing more willing to heare such things told, than bolde to attempt the triall, told him that he was not willing to see any such experiment.

"And by this it appeareth, that all the *Negroes* are addicted to Incantations, which never have anie approbation from God, except against Serpents, which I cannot very easilie be brought to beleeve."

Of the affection of some serpents for the human-kind he gives some examples :—" We reade also in Plutarch of certain Serpents, lovers of young virgins, and by name there was one that was in love with one *Ætolia*, a Virgin, who did accustome to come unto her in the night time, slyding gentlie all over her bodie, never harming her, but as one glad of such acquaintance, tarried with her in that dalliance till the morning, and them would depart away of his owne accorde: the which thing beeing made manifest unto the Guardians and Tutours of the Virgin, they removed her unto another Towne. The Serpent missing his Love, sought her uppe and downe three or four dayes, and at last mette her by chance, and then hee saluted her not as he was wont, with fawning, and gentle slyding, but fiercely assaulted her with grimme and austere countenance, flying to her hands, and binding them with the spire of his bodie, fast to her sides, did softly with his tayle beat her upon her backer parts. Whereby was collected, some token of his chastisement unto her, who had wronged such a Lover, with her wilfull absence and disappointment.

"It is also reported by *Ælianus* that *Egemon* in his verses, writeth of one *Alena*, a *Thessalian* who, feeding his Oxen in *Thessaly*, neere the Fountaine *Hæmonius*, there fell in love with him a Serpent of exceeding bignesse and quantitie, and the same would come unto him, and softly licke his face and golden haire, without dooing him any manner of hurt at all."

He tells a few more "Snake stories," and quotes from "a little Latine booke printed at *Vienna*, in the yeare of the Lorde 1551," the following :—" There was (sayth mine Author) found in a mowe or rycke of corne, almost as many Snakes, Adders, and other Serpentes, as there were sheafes, so as no one sheafe could be removed, but there presently appeared a heape of ougly and fierce Serpents. The countrey men determined to set fire upon the Barne, and so attempted to doe, but in vaine, for the straw would take no fire, although they laboured with all their wit and pollicye, to burne them up; At last, there appeared unto them at the top of the heap a huge great Serpent, which, lifting up his head, spake with man's voyce to the countrey men, saying : *Cease to prosecute your devise, for you shall not be able to accomplish our burning, for wee were not bredde by Nature, neither came we hither of our own accord, but were sent by God to take vengeance on the sinnes of men.*"

And some serpents were "very fine and large," for he says :—" *Gellius* writeth, that when the Romanes were in the Carthaginian Warre, and *Attilius Regulus* the Consull had pitched his Tents neere unto the river *Bragrada*, there was a Serpent of monstrous quantitie, which had beene lodged within the compasse of the Tents, and therefore did cause to the whole Armie exceeding great calamitie, untill by casting of stones with slings, and many other devises, they oppressed and slew that Serpent, and afterward fleyed off the skinne and sent it to *Rome;* which was in length one hundred and twentie feete.

"And, although this seemeth to be a beast of unmatchable stature, yet *Posidenius* a Christian writer, relateth a storie of another which was much greater, for hee writeth that he saw a Serpent dead, of the length of an acre of Land, and all the residue both of head and bodie, were answerable in proportion, for the bulke of his bodie was so great, and lay so high, that two Horsemen could not see one the other, beeing at his two sides, and the widenes of his mouth was so great, that he could receive at one time, within the compasse thereof, a horse and a man on his backe both together : The scales of his coate or skinne, being every one like a large buckler or target. So that now, there is no such cause to wonder at the Serpent which is said to be killed by *St. George*, which was, as is reported, so great, that eight Oxen were but strength enough to drawe him out of the Cittie *Silena*. . . .

"Among the *Scyritæ*, the Serpents come by great swarmes uppon their flocks of sheepe and cattell, and some they eate up all, others they kill, and sucke out the blood, and some part they carry away. But if ever there were anything beyond credite, it is the relation of *Volateran* in his twelfth booke of the *New-found Lands*, wherein he writeth, that there are Serpents of a mile long, which at one certaine time of the yeere come abroad out of the holes and dennes of habitation, and destroy both the Heards and Heard-men if they find them. Much more favourable are the Serpents of a *Spanish* Island, who doe no harme to any living thing, although they have huge bodies, and great strength to accomplish their desires."

After this it will be refreshing to have one of Topsell's own particular *true stories :* and this is " Of a true history done in *England*, in the house of a worshipfull Gentleman, upon a servant of his, whom I could name if it were needfull. He had a servant that grew very lame and feeble in his legges, and thinking that he could never be warme in his bed, did multiply his clothes, and covered himselfe more and more, but all in vaine, till at length he was not able to goe about, neither could any skill of Phisitian or Surgeon find out the cause.

"It hapned on a day as his Maister leaned at his Parlour window, he saw a great Snake to slide along the house side, and to creepe into the chamber of this lame man, then lying in his bedde, (as I remember,) for hee lay in a lowe chamber, directly against the Parlour window aforesaid. The Gentleman desirous to see the issue, and what the Snake would doe in the chamber, followed, and looked into the chamber by the window; where hee espied the snake to slide uppe into the bed-straw, by some way open in the bottome of the bedde, which was of old bordes. Straightway, his hart rising thereat, he called two or three of his servaunts, and told them what he had seene, bidding them goe take their Rapiers, and kill the said snake. The serving-men came first, and removed the lame man (as I remember) and then the one of them turned up the bed, and the other two the straw, their Maister standing without, at the hole, whereinto the said snake had entered into the chamber. The bedde was no sooner turned up, and the Rapier thrust into the straw, but there issued forth five or six great snakes that were lodged therein: Then the serving-men bestirring themselves, soone dispatched them, and cast them out of doores dead. Afterward, the lame man's legges recovered, and became as strong as ever they were; whereby did evidentlie appeare, the coldnes of these snakes or Serpents, which came close to his legges everie night, did so benumme them, as he could not goe."

Yet one more :—

"I cannot conceale a most memorable historie as ever was any in the world, of a fight betwixt the Serpents of the Land and the Water. This history is taken out of a Booke of *Schilt-bergerus*, a *Bavarian*, who knew the same, (as he writeth) while hee was a captive in *Turky;* his words are these. In the kingdome called *Genyke,* there is a Citty called *Sampson*, about which, while I was prisoner with *Baiazeta* King of *Turkes,* there pitched or arrived, an innumerable company of Land and Water Serpents, compassing the said Cittie, a mile about. The Land Serpents came out of the woods of *Tricnick,* which are great and many, and the Water Serpents came out of the bordering Sea. These were nine dayes together assembling in that place, and for feare of them there was not any man that durst goe out of the Citty, although it was not observed that they hurt any man, or living creature there-abouts.

"Wherefore the Prince also commanded, that no man should trouble them, or doe them any harme, wisely judging, that such an accident came not but by Divine Miracle, and that also to signifie some notable event. Uppon the tenth day, these two valiant troupes joyned battell, early in the morning, before the sunne-rising, so continuing in fight untill the sunne-set, at which time the Prince, with some horsemen, went out of the Cittie to see the battell, and it appeared to him and his associates, that the Water Serpents gave place to the Land Serpents. So the Prince, and his company, returned into the Citty againe, and the next day went forth againe, but found not a Serpent alive, for there were slaine above eyght thousand: all which, he caused presently to be covered with earth in ditches, and afterwards declared the whole matter to *Baiazeta* by letters, after he had gotten that Cittie, whereat the great Turke rejoyced, for hee thereby interpreted happinesse to himselfe."

Luckily, man has found out things inimical to Serpents, and they, and their use, seem to be very simple:—

"There is such vertue in the Ashe tree, that no Serpent will endure to come neere either the morning or evening shadow of it; yea, though very farre distant from them, they do so deadlie hate it. We set downe nothing but that wee have found true by experience: If a great fire be made, and the same fire encircled round with Ashen-boughes, and a serpent put betwixt the fire and the Ashen-boughes, the Serpent will sooner runne into the fire, than come neere the Ashen-boughes: thus saith *Pliny*. *Olaus Magnus* saith, that those Northern Countries which have great store of Ash-trees, doe want venemous beasts, of which opinion is also *Pliny*. *Callimachus* saith, there is a Tree growing in the Land of *Trachinia*, called *Smilo*, to which, if any Serpents doe either come neere, or touch, they foorthwith die. *Democritus* is of opinion, that any Serpent will die if you cast Oken-leaves upon him. *Pliny* is of opinion that *Alcibiadum*, which is a kind of wild Buglosse, is of the same use and qualitie; and further, being chewed, if it be spet upon any serpent, that it cannot possibly live. In time of those solemne Feastes which the *Athenians* dedicated to the Goddesse *Ceres*, their women did use to lay and strew their beddes, with the leaves of the Plant called *Agnos*, because serpents could not endure it, and because they imagined it kept them chast, Where-upon they thought the name was given it. The herbe called Rosemarie, is terrible to serpents.

"The *Egyptians* doe give it out, that *Polydamna* the wife of *Thorris* their King, taking pittie upon *Helen*, caused her to be set on shore in the Island of *Pharus*, and bestowed upon her an herbe (whereof there was plenty) that was a great enemy to serpents: whereof the serpents having a feeling sence (as they say) and so readily knowne of them, they straightwaies got them to their lurking holes in the earth; and *Helen* planted this herbe, who, coming to the knowledge thereof, she perceived that in his due time it bore a seede that was a great enemy to serpents, and thereupon was called *Helenium*, as they that are skilfull in Plants affirme; and it groweth plentifully in *Pharus*, which is a little Ile against the mouth of *Nylus*, joyned to *Alexandria* by a bridge.

"Rue, (called of some, Herbe of Grace) especially that which groweth in *Lybia*, is but a backe friend to Serpents, for it is most dry, and therefore causing Serpents soon to faint, and loose their courage, because (as *Simocatus* affirmeth) it induceth a kind of heavinesse or drunkennesse in their head, with a vertiginie, or giddines through the excesse of his drinesse, or immoderate sticcitie. Serpents cannot endure the savour of Rue, and, therefore, a Wesill, when she is to fight with any serpent, eateth Rue, as a defensative against her enemie, as *Aristotle*, and *Pliny* his Interpreter, are of opinion.

"The Country people leaving their vessels of Milke abroade in the open fieldes, doe besmeare them round about with garlick, lest some venomous serpents should creepe into them, but the smell of garlick, as *Erasmus* saith, driveth them away. No serpents were ever yet seene to touch the herbe *Trifolie*, or Three-leaved-grasse, as *Ædonnus* wold make us believe. And *Cardan* the Phisitian hath observed as much, that serpents, nor anything that is venemous will neither lodge, dwell, or lurk privily neere unto *Trifolie*, because that is their bane, as they are to other living creatures: and therefore it is sowne to very good purpose, and planted in very hot countries, where there is most store of such venomous creatures.

"*Arnoldus Villanonanus* saith that the herb called *Dracontea* killeth serpents. And *Florentinus* affirmeth that, if you plant Woormwood, Mugwort, or Sothernwood about your dwelling, that no venomous serpents will ever come neer, or dare enterprise to invade the same. No serpent is found in Vines, when they flourish, bearing flowers or blossoms, for they abhor the smell, as *Aristotle* saith. *Avicen*, an *Arabian* Phisitian, saith, that Capers doe kill worms in the guts, and likewise serpents. If you make a round circle with herbe Betonie, and therein include any serpents, they will kill themselves in the place, rather than strive to get away. Galbanum killeth serpents only by touching, if oyle and the herbe called Fenell-giant be mixt withall. There is a shrubbe called Therionarca, having a flower like a Rose, which maketh serpents heavy, dull and drousie, and so killeth them, as *Pliny* affirmeth."

There are more plants inimical to serpents, but enough have been given to enable the reader, if he have faith in them, to defend himself; and it is comforting to think, that although the serpent is especially noxious, when alive, he is marvellously useful, medicinally, when dead.

Even now, in some country places, viper broth is used as a medicine; and, in the first half of the eighteenth century, its flesh, prepared in various ways, was thoroughly recognised in the Pharmacopœia. But Topsell, who gathered together all the wisdom of the ancients, gives so very many remedies (for all kinds of illnesses) that may be derived from different parts, and treatment, of serpents, that I can only pick out a few:—

"*Pliny* saith, that if you take out the right eye of a serpent, and so bind it about any part of you, that it is of great force against the watering or dropping of the eyes, by meanes of a rhume issuing out thereat, if the serpent be againe let goe alive. And so hee saith, that a serpent's or snake's hart, if either it be bitten or tyed to any part of you, that it is a present remedie for the toothach: and hee addeth further, that if any man doe tast of the snake's hart, that he shall never after be hurt of any serpent. . . . The blood of a serpent is more precious than *Balsamum*, and if you annoynt your lips with a little of it, they will looke passing redde: and, if the face be annoynted therewith, it will receive no spot or fleck, but causeth it to have an orient and beautiful hue. It represseth all scabbiness of the body, stinking in the teeth, and gummes, if they be therewith annointed. The fat of a serpent speedily helpeth all rednes, spots, and other infirmities of the eyes, and beeing annoynted upon the eyeliddes, it cleereth the eyes exceedingly.

"Item, put them (*serpents*) into a glassed pot, and fill the same with Butter in the Month of May, then lute it well with paste (that is, Meal well kneaded) so that nothing may evaporate, then sette the potte on the fire, and let it boyle wel-nigh halfe a day: after this is done, straine the Butter through a cloth, and the remainder beate in a morter, and straine it againe, and mixe them together, then put them into water to coole, and so reserve it in silver or golden boxes, that which is not evaporated, for the older, the better it is, and so much the better it will be, if you can keepe it fortie years. Let the sicke patient, who is troubled eyther with the Goute, or the Palsie, but annoynt himselfe often against the fire with this unguent, and, without doubt, he shall be freed, especially if it be the Goute."

Of serpents in general, I shall have little to say, except those few of which the descriptions are the most *outré*. And first let us have out the "Boas," which cannot mean that enormous serpent the Boa-Constrictor, which enfolds oxen, deer, &c., crushing their bones in its all-powerful fold, and which sometimes reaches the length of thirty or five-and-thirty feet—long enough, in all conscience, for a respectable serpent. But Topsell begins his account of " The Boas " far more magnificently :—

" It was well knowne among all the Romans, that when *Regulus* was Governour, or Generall, in the *Punick* warres, there was a Serpent (neere the river *Bagrade*) killed with slings and stones, even as a Towne or little Cittie is over-come, which Serpent was an hundred and twenty foote in length ; whose skinne and cheeke bones, were reserved in a Temple at *Rome*, untill the *Numantine* warre.

"And this History is more easie to be beleeved, because of the Boas Serpent bred in Italy at this day : for we read in *Solinus*, that when *Claudius* was Emperour, there was one of them slaine in the *Vatican* at Rome, in whose belly was found an Infant swallowed whole, and not a bone thereof broken. . . .

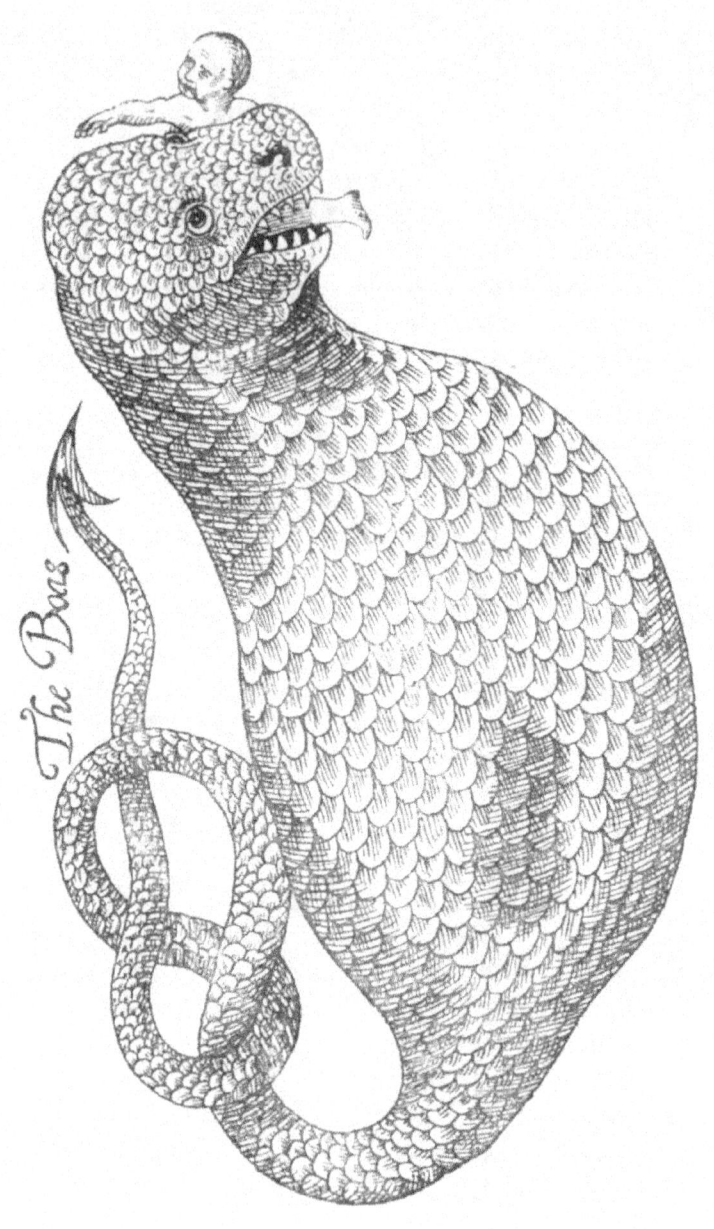

"The Latines call it *Boa*, and *Bova*, because by sucking Cowe's milke it so encreaseth, that in the end it destroyeth all manner of herdes, Cattell, and Regions. . . . The Italians doe usually call them, *Serpeda de Aqua*, a Serpent of the water, and, therefore, all the Learned expound the Greeke word *Hydra*, for a Boas. *Cardan* saith, that there are of this kind in the Kingdom of *Senega*, both without feet and wings, but most properly, as they are now found in Italy, according to these verses:

> *Boa quidem serpens quem tellus Itala nutrit*
> *Hunc bubulum plures lac enutrire docent.*

Which may be englished thus:

> *The Boas Serpent which Italy doth breede,*
> *Men say, uppon the milke of Cowes doth feede.*

"Their fashion is in seeking for their prey among the heardes, to destroy nothing that giveth suck, so long as it will live, but they reserve it alive untill the milk be dryed up, then afterwards they kill and eate it, and so they deale with whole flocks and heards."

Whilst on the subject of Hydra, I give Topsell's idea of the Lernean Hydra, whose story is so familiar to us. (See p. 292.) But, after presenting us with such a frightful ideal, he says:—" And some ignorant men of late daies at *Venice*, did picture this Hydra with wonderfull Art, and set it forth to the people to be seene, as though it had beene a true carkase, with this inscription: In the yeare of Christe's incarnation, 550, about the Month of January, 'this monstrous Serpent was brought out of *Turky* to *Venice*, and afterwards given to the French King: It was esteemed to be worth 600 duckats. These monsters

signifie the mutation or change of worldly affaires,' &c."
And, after giving a long-winded inscription, *àpropos* of
nothing, he says : —" I have also heard that in *Venice* in

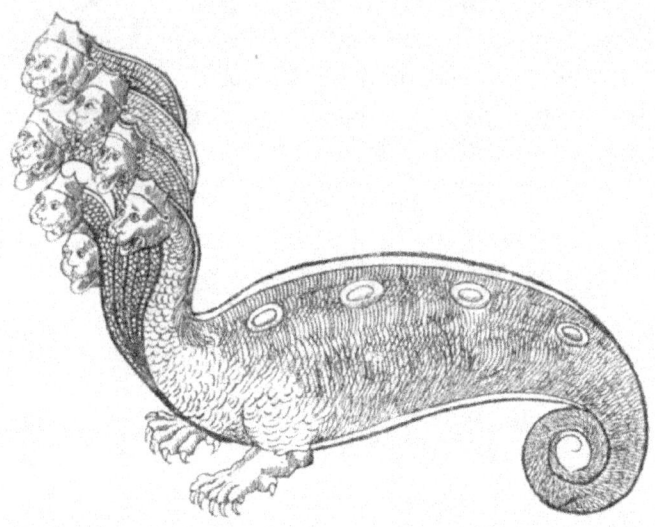

the Duke's treasury, among the rare Monuments of that
Citty, there is preserved a Serpent with seaven heads,
which, if it be true, it is the more probable that there is a Hydra, and that the Poets were not altogether deceived, that say *Hercules* killed such an one."

Mr. Henry Lee, in his little book, "Sea Fables Explained," says that the Lernean Hydra was neither more nor less than a huge Octopus, and gives an illustration of a marble tablet in the Vatican (also given in

"Smith's Classical Dictionary"), which does not seem unlike one.

The Wingless Dragons belong to the serpent tribe, with the exception that they are generally furnished with legs. These are "Wormes," of several of which we, in England, were the happy possessors. Of course, in the northern parts of Europe, they survived (in story at all events) much later than with us, and Olaus Magnus gives accounts of several fights with them, notably that of Frotho and Fridlevus, two Champions, against a serpent.

"*Frotho*, a Danish Champion and a King, scarce being past his childhood, in a single combat killed a huge fierce great Serpent, thrusting his sword into his belly, for his hard skin would not be wounded, and all darts

thrown at him, flew back again, and it was but labour lost. *Fridlevus* was no lesse valiant, who, both to try his valour, and to find out some hidden treasure, set upon a most formidable Serpent for his huge body and venomous teeth, and, for a long time, he cast his darts against his scaly sides, and could not hurt him, for his hard body made nothing of the weapons cast with violence against him. But this Serpent twisting his tail in many twines, by turning his tail round, he would pull up trees by the roots, and by his crawling on the ground, he had made a great hollow place, that in some places, hills seemed to be parted as if a valley were between them, wherefore *Fridlevus* considering that the upper parts of this beast could not be penetrated, he runs him in with his sword underneath; and, piercing into his groine, he drew forth his virulent matter, as he lay panting: when he had killed the Serpent, he dug up the money, and carried it away."

He gives another story of a combat with "Wormes," although in the Latin they are called *Vipers:* yet I leave my readers to judge whether the small snake, the viper, would require such an amount of killing as Regner had to bestow upon them :—

"Of *Regnerus,* called Hair-Coat. There was a King of the *Sueons* called *Herothus,* whose troubled mind was not a little urged how to preserve his Daughter's chastity; whether he should guard her with wild beasts (as the manner of most Princes was then) or else should commit the custody of her to man's fidelity. But he, preferring cruelty of Beasts to man's fidelity, he soonest chose what would do most hurt. For, hunting in the woods, he brought some Snakes that his Company had found, for his Daughter to feed up. She, quickly obeying her Father's commands, bred up a generation of vipers by her Virgin hands. And that they might want no meat, her curious Father caused the whole body of an Ox to be brought, being ignorant that, by this private food, he maintain'd a publick destruction. These, being grown up, by their venomous breath poysoned the neighbouring parts; but the King, repenting his folly, proclaimed that he who could remove this plague, should have his daughter.

"When *Regnerus* of *Norway*, descended of the King's race, who was the chief Suiter this Virgin had, heard this Report, he obtained from the Nurse a woollen Cassock, and hairy Breeches, whereby he might hinder the biting of the Adders. And when he came to *Sweden* in a ship, he purposely suffered his Clothes to grow stiff with cold, casting water upon them: and thus clothed, having onely his Sword and Dart to defend him, he went to the King. As he went forward, two huge Adders met him on the way, that would kill the young man, with the twisting of their tails, and by the venome they cast forth.

"But *Regnerus* confiding in the hardness of his frozen Garments, both endured and repulsed their Venome, by his clothes, and their biting his Harness, being indefatigable in pressing hard upon these Wild Beasts. Last of all he strongly casts out of his hand his Javelin that was fastened with a Hoop, and struck it into their bodies. Then, with his two-edged Sword, rending both their hearts, he obtained a happy end of an ingenious and dangerous fight. The King, looking curiously on his clothes, when he saw them so hairy on the back-side, and unpolished like ragged Frize, he spake merrily, and called him *Lodbrock:* that is *Hair Coat;* and to recreate him after his pains, he sent for him to a Banquet with his friends. He answered, *That he must first go see those Companions he had left:* and he brought them to the King's Table, very brave in clothes, as he was then: and lastly, when that was done, he received the pledge of his Victory, by whom he begat many hopeful Children: and he had her true love to him the more, and the rather enjoyed his company, by how much she knew the great dangers he underwent to win her by, and the ingenious practises he used."

We were favoured in England with several "Wormes." Nor only in England, but in Scotland and Wales. Of course, Ireland can boast of none, as St. Patrick banished all the serpents from that island.

Of the Dragon of Wantley I say nothing; he has been reslain in modern times, and all the romance has gone out of him. Nobody wishes to know that the Dragon was Sir Francis Wortley, who was at loggerheads with his neighbours, notably one Lionel Rowlestone, whose advocate was More of More Hall. We had rather have had our dear old Dragon, and have let the champion More slay him in the orthodox manner.

But the "laidley Worme" of Lambton is still all our own, and its story is thus told by Surtees in his "History, &c., of Durham," 1820:—

"The heir of Lambton, fishing, as was his profane custom, in the Wear, on a Sunday, hooked a small worm or eft, which he carelessly threw into a well, and thought no more of the adventure. The worm (at first neglected) grew till it was too large for its first habitation, and, issuing forth from the *Worm Well*, betook itself to the Wear, where it usually lay a part of the day coiled round a crag in the middle of the water; it also frequented a green mound near the well (*the Worm Hill*), where it lapped itself nine times round, leaving vermicular traces, of which, grave living witnesses depose that they have seen the vestiges. It now became the terror of the country, and, amongst other enormities, levied a daily contribution of nine cows' milk, which was always placed for it at the green hill, and in default of which it devoured man and beast. Young Lambton had, it seems, meanwhile, totally repented him of his former life and conversation, had bathed himself in a bath of holy water, taken the sign of the cross, and joined the Crusaders.

"On his return home, he was extremely shocked at witnessing the effects of his youthful imprudences, and immediately undertook the adventure. After several fierce combats, in which the Crusader was foiled by his enemy's *power of self-union*, he found it expedient to add policy to courage, and not, perhaps, possessing much of the former quality, he went to consult a witch or wise woman. By her judicious advice he armed himself in a coat-of-mail studded with razor blades; and, thus prepared, placed himself on the crag in the river, and awaited the monster's arrival.

"At the usual time the worm came to the rock, and wound himself with great fury round the armed knight, who had the satisfaction to see his enemy cut in pieces by his own efforts, whilst the stream washing away the severed parts, prevented the possibility of reunion.

"There is still a sequel to the story: the witch had promised Lambton success only on one condition, that he should slay the first living thing which met his sight after the victory. To avoid the possibility of human slaughter, Lambton had directed his father, that as soon as he heard him sound three blasts on his bugle, in token of the achievement performed, he should release his favourite greyhound, which would immediately fly to the sound of the horn, and was destined to be the sacrifice. On hearing his son's bugle, however, the old chief was so overjoyed, that he forgot his instructions, and ran himself with open arms to meet his son. Instead of committing a parricide, the conqueror again repaired to his adviser, who pronounced, as the alternative of disobeying the original instructions, that no chief of the Lambtons should die in his bed for seven, (or as some accounts say) for nine generations—a commutation which, to a martial spirit, had nothing probably very terrible, and which was willingly complied with. . . .

"In the garden-house at Lambton are two figures of no great antiquity. A Knight in good style, armed cap-a-pie, the back *studded with razor blades*, who holds the worm by one ear with his left hand, and with his right crams his sword to the hilt down his throat; and a Lady who wears a coronet, with bare breasts, &c., in the style of Charles 2nd's Beauties, a wound on whose bosom and an accidental mutilation of the hand are said to have been the work of the worm."

There were several other English "Wormes," but this must suffice as a type. Also, as a typical Scotch "Worme," the Linton Worme will serve. A writer (W. E.) tells its story so well in *Notes and Queries*, February 24, 1866, that I transfer it here, in preference to telling it myself. It was slain by Sir John Somerville, about the year 1174, who received the lands and barony of Linton, in Roxburghshire, as the reward of his exploit. W. E. quotes from a family history entitled a "Memorie of the Somervills," written by James, the eleventh lord, A.D. 1679 :—

"'In the parochene of Lintoune, within the sheriffdome of Roxburghe, ther happened to breede ane hydeous monster, in the forme of a worme, soe called and esteemed by the country people (but in effecte has beene a serpente or some suche other creature), in length three

Scots yards, and somewhat bigger than ane ordinarie man's leg, &c. . . . This creature, being a terrour to the country people, had its den in a hollow piece of ground, on the syde of a hill, south east from Lintoun Church, some more than a myle, which unto this day is knowne by the name of the Worme's glen, where it used to rest and shelter itself; but, when it sought after prey, then would it wander a myle or two from its residence, and make prey of all sort of bestiall that came in its way, which it easily did because of its lownesse, creeping amongst the peat, heather, or grasse, wherein that place abounded much, by reason of the meadow grounde, and a large flow moss, fit for the pasturage of many cattell. . . . Soe that the whole country men thereabout wer forced to remove ther bestiall and transport them 3 or 4 myles from the place, leaving the country desolate, neither durst any person goe to the Church, or mercat, upon that rod, for fear of this beast.'

"Somerville happening to come to Jedburgh, on the King's business, found the inhabitants full of stories about the wonderful beast.

"'The people who had fled ther for shelter, told soe many lies, as first, that it increased every day, and was beginning to get wings: others pretended to have seen it in the night, and asserted it was full of fyre, and in tyme, would throw it out, &c., with a thousand other ridiculous stories.'

"Somerville determined to see the monster, and, accordingly, rode to the glen about sunrise, when he was told it generally came forth. He had not to wait long, till he perceived it crawl out of its den. When it observed him, it raised itself up, and stared at him, for some time, without venturing to approach; whereupon he drew nearer to observe it more closely, on which it turned round, and slunk into its lair.

"Satisfied that the beast was not so dangerous as reported, he resolved to destroy it, but as every one declared that neither sword nor dagger had any effect on it, and that its venom would destroy any one that came within its reach: he prepared a spear double the ordinary length, plated with iron, four feet from the point, on which he placed a slender iron wheel, turning on its centre. On this he fastened a lighted peat, and exercised his horse with it for several days, until it shewed no fear or dislike to the fire and smoke. He then repaired to the den, and, on the worme appearing, his servant set fire to the peat, and, putting spurs to his horse, he rode full at the beast. The speed at which he advanced, caused the wheel to spin round, and fanned the peat into a blaze. He drove the lance down the monster's throat full a third part of its length, when it broke, and he left the animal writhing in the agonies of death."

I am afraid the Welsh "Worme" is not so well authenticated as the others; but the story is, that Denbigh is so named from a Dragon slain by John Salusbury of Lleweni, who died 1289. It devastated the country far and wide, after the manner of its kind, and all the inhabitants prayed for the destruction of this *bych*. This the Champion effected, and in his glee, joyfully sang, *Dyn bych, Dyn bych* (*No bych*); and the country round was so named.

There arises the question, whether, having regard to the fact that the Lambton worm, at all events, was amphibious, it might not have been a Plesiosaurus, which had survived some of its race, such as the illustration now given, of the one reconstructed by Thos. Hawkins, in his "Book of the Great Sea Dragons." We know that at some time or other these animals existed, and, it may be, some few lingered on. At all events most civilised nations have had a belief in it, and it was held to be the type of all that was wicked; so much so, that one of Satan's synonyms is "the Great

Dragon." In the Romances of Chivalry, its destruction was always reserved for the worthiest knight; in classical times it was a terror. Both Hindoos and Chinese hold it in firm faith, and, take it all in all, belief in its entity was general.

The Winged Dragons were undoubtedly more furious and wicked than the Wormes, and there is scarcely any reason to go farther than its portrait by Aldrovandus, to enable us to recognise it at any time. (See next page.) Topsell gives another, but with scarcely so much detail.

But, although we in our times have not seen flying dragons in the flesh, we have their fossilised bones in evidence of their existence. The Pterodactyl, as Mr. Hawkins observes, "agrees with the Dragon in nearly all its more important features. Thus, it was of great size, possessed a large head, with long jaws and powerful teeth. It had wings of great span, and at the same time three powerful clawed fingers to each hand, wings devoid of feathers, and capable of being folded along the sides of the body, while the large size of the orbits may not, improbably, have suggested the name dragon; for dragon, which is derived from the Greek δράκων, means, literally, *keen-sighted*."

We now have flying lizards, both in India and the Malay Archipelago, in which latter is found a small lemur which can fly from tree to tree, and we are all familiar with bats, some of which attain a large size.

Topsell has exercised great research among old authorities respecting dragons, and he draws their portraits thus:—" *Gyllius*, *Pierius*, and *Grevinus*, following the authority of *Nicander*, do affirme that a Dragon is of a blacke colour, the bellie somewhat green, and very beautifull to behold, having a treble rowe of teeth in their mouthes upon every jawe, and with most bright and cleare seeing eyes, which caused the Poets to faine in their writings, that these dragons are the watchfull keepers of Treasures. They have also two dewlappes growing under their chinne, and hanging downe like a beard, which are of a redde colour; their bodies are set all over with very sharpe scales, and over their eyes

stand certaine flexible eyeliddes. When they gape wide with their mouth, and thrust forth their tongue, theyr teeth seeme very much to resemble the teeth of Wilde Swine: And theyr neckes have many times grosse thicke hayre growing upon them, much like unto the bristles of a Wylde Boare."

Apart from looks, he does not give dragons, as a rule, a very bad character, and says they do not attack men unless their general food fails them :—"They greatlie preserve their health (as *Aristotle* affirmeth) by eating of Wild lettice, for that they make them to vomit, and cast foorth of theyr stomacke what soever meate offendeth them, and they are most speciallie offended by eating Apples, for theyr bodies are much subject to be filled with winde, and therefore they never eate Apples, but first they eate Wilde lettice. Theyr sight also (as *Plutarch* sayth) doth many times grow weake and feeble, and therefore they renew and recover the same againe by rubbing their eyes against Fennel, or else by eating it. Their age could never yet be certainely knowne, but it is conjectured that they live long, and in great health, like all other serpents, and therefore they grow so great.

. . . .

"Neither have wee in Europe onely heard of Dragons, and never seene them, but also even in our own Country, there have (by the testimonie of sundry writers) divers been discovered and killed. And first of all, there was a Dragon, or winged Serpent, brought unto *Francis* the French King, when hee lay at *Sancton*, by a certaine Country man, who had slaine the same Serpent himselfe with a Spade, when it sette upon him in the fields to kill him. And this thinge was witnessed by many

Learned and Credible men which saw the same; and they thought it was not bredde in that Country, but rather driven by the winde thither from some forraine Nation. For Fraunce was never knowne to breede any such Monsters. Among the *Pyrenes*, too, there is a cruell kinde of Serpent, not past foure foot long, and as thicke as a man's arme, out of whose sides growe winges, much like unto gristles.

"*Gesner* also saith, that in the yeere of our Lord 1543 there came many Serpents both with wings and legs into the parts of Germany neere *Stiria*, who did bite and wound many men incurably. *Cardan* also describeth certaine serpents with wings, which he saw at Paris, whose dead bodies were in the hands of *Gulielmus Musicus;* hee saith that they had two legges, and small winges, so that they could scarce flie, the head was little, and like to the head of a serpent, their colour bright, and without haire or feathers, the quantitie of that which was greatest, did not exceede the bignes of a Cony, and it is saide they were brought out of India. . . .

"There have beene also Dragons many times seene in Germaine, flying in the ayre at mid-day, and signifying great and fearefull fiers to follow, as it happened neere to the Cittie called *Niderburge*, neere to the shore of the *Rhyne*, in a marvailous cleere sun-shine day, there came a dragon three times successively together in one day, and did hang in the ayre over a Towne called *Sanctogoarin*, and shaking his tayle over that Towne every time: it appeared visibly in the sight of many of the inhabitants, and, afterwards it came to passe, that the said towne was three times burned with fire, to the great harme and undooing of the people dwelling in the same; for they were not able to make any resistance to quench the fire, with all the might, Art, and power they could raise. And it was further observed, that about the time there were many dragons seene washing themselves in a certaine Fountaine or Well neere the towne, and if any of the people did by chance drinke of the water of that Well, theyr bellyes did instantly begin to swell, and they dyed as if they had been poysoned. Whereupon it was publicly decreed, that the said well should be filled up with stones, to the intent that never any man should afterwards be poisoned with that water; and so a memory thereof was continued, and these thinges are written by *Justinus Goblerus*, in an Epistle to *Gesner*, affirming that he did not write fayned things, but such things as were true, and as he had learned from men of great honesty and credite, whose eyes did see and behold both the dragons, and the mishaps that followed by fire."

Hitherto we have only seen that side of a Dragon's temperament that is inimical to man, but there are stories, equally veracious, of their affection and love for men, women, and children: how they, by kindness, may be tamed, and brought into kindly relations with the human species.

Pliny, quoting *Democritus*, says that "a Man, called *Thoas*, was preserved in *Arcadia* by a Dragon. When a boy he had become much attached to it, and had reared it very tenderly; but his father, being alarmed at the nature and monstrous size of the reptile, had taken and left it in the desert. *Thoas* being here attacked by some robbers, who lay in ambush, he was delivered from them by the Dragon, which recognised his voice, and came to his assistance."

Topsell tells us that "there be some which by certaine inchaunting verses doe tame Dragons, and rydeth upon their neckes, as a man would ride upon a horse, guiding and governing them with a bridle."

And so widely spread was the belief that these fearful animals could be brought into subjection, that Magnus gives us an account "Of the Fight of King *Harald* against a tame Dragon," but this one seems hardly as docile as those previously instanced:—"*Haraldus* the most illustrious King of *Norway*, residing, in his youth, with the King of *Constantinople*, and being condemned

for man-slaughter, he was commanded to be cast to a tame Dragon that should rend him in pieces. As he went into the prison, one very faithfull servant he had, offered himself freely to die with his Master.

"The keeper of the Castle, curiously observing them both, let them down at the mouth of the Den, being unarmed, and well searched; wherefore, when the servant was naked, he admitted *Harald* to be covered with his shirt, for modesty's sake, who gave him a braslet privily, and he scattered little fish on the pavement, that the Dragon might first stay his hunger on them, and that the guilty persons that are shut up in the dark prison, might have a little light by the shining of the Fins and Scales. Then *Haraldus* picking up the bones of a Carkeis, stopt them into the linen he had, and bound them fast together like a Club. And when the Dragon was let forth, and rushed greedily on his prey cast to him, he lept quickly on his back, and he thrust a Barber's razor in at his navill, that would only be pierced by iron, which, as luck was, he brought with him, and kept it concealed by him: this cold Serpent that had most hard scales all over, disdained to be entred in any other part of his body. But *Haraldus* sitting so high above him, could neither be bitten by his mouth, or hurt by his sharp teeth; or broken with the turnings of his tayle. And his servant using the weapons, or bones put together, beat the Dragon's head till he bled, and died thereof by his many weighty strokes. When the King knew this, he freely changed his revenge, into his service, and pardoned these valiant persons, and furnishing them with a Ship and Monies, he gave them leave to depart."

The natural instinct of Dragons was undoubtedly vicious, and they must have been most undesirable neighbours, *teste* the following story quoted by Topsell from Stumpsius:—"When the Region of *Helvetia* beganne first to be purged from noysome beasts, there was a horrible dragon found neere a Country towne called *Wilser*, who did destroy all men and beastes, that came within his danger in the time of his hunger, inasmuch that that towne and the fields therto adjoyning, was called *Dedwiler*, that is, a Village of the Wildernes, for all the people and inhabitants had forsaken the same, and fledde to other places.

"There was a man of that Towne whose name was *Winckleriedt*, who was banished for manslaughter: this man promised, if he might have his pardon, and be restored againe to his former inheritance, that he would combat with that Dragon, and by God's helpe destroy him; which thing was granted unto him with great joyfulnes. Wherefore he was recalled home, and in the presence of many people went foorth to fight with that Dragon, whom he slew and overcame, whereat for joy hee lifted uppe his sword imbrued in the dragon's blood, in token of victory, but the blood distilled downe from the sword uppon his body, and caused him instantly to fall downe dead."

"There be certaine beasts called *Dracontopides*, very great and potent Serpents, whose faces are like to the faces of Virgins, and the residue of their body like to dragons. It is thought that such a one was the Serpent that deceived *Eve*, for *Beda* saith it had a Virgin's countenance, and therefore the woman, seeing the likenes of her owne face, was the more easily drawne to believe it: into which the devill had entred; they say he taught it to cover the body with leaves, and to shew nothing but the head and face. But this fable is not worthy to be refuted, because the Scripture itself, dooth directly gaine-say everie part of it. For, first of all it is called a Serpent, and if it had been a Dragon, *Moses* would have said so; and, therefore, for ordinary punishment, GOD doth appoint it to creepe upon the belly, wherefore it is not likely that it had either wings or feete. Secondly, it was impossible and unlikely, that any part of the body was covered or conceiled from the sight of the woman, seeing she knew it directly to be a Serpent, as shee afterward confessed before GOD and her husband.

"There be also certaine little dragons called in *Arabia, Vesga*, and in *Catalonia, Dragons of houses;* these, when they bite, leave their teeth behind them, so as the wound never ceaseth swelling, as long as the teeth remain therein, and therefore, for the better cure thereof, the teeth are drawne forth, and so the wound will soone be healed.

"And thus much for the hatred betwixt men and dragons, now we will proceed to other creatures.

"The greatest discord is between the Eagle and the Dragon, for the Vultures, Eagles, Swannes and Dragons, are enemies to one another. The Eagles, when they shake their winges, make the dragons afraide with their ratling noyse; then the dragon hideth himselfe within his den, so that he never fighteth but in the ayre, eyther when the Eagle hath taken away his young ones, and he, to recover them, flieth aloft after her, or else when the Eagle meeteth him in her nest, destroying her egges and young ones: for the Eagle devoureth the dragons, and little Serpents upon earth, and the dragons againe, and Serpents do the like against the Eagles in the ayre. Yea, many times the dragon attempteth to take away the prey out of the Eagle's talants, both on the ground, and in the ayre, so that there ariseth betwixt them a very hard and dangerous fight.

"In the next place we are to consider the enmitie that is betwixt Dragons and Elephants, for, so great is their hatred one to another, that in Ethyopia the greatest dragons have no other name but Elephant killers. Among the Indians, also, the same hatred remaineth, against whom the dragons have many subtile inventions: for, besides the greate length of their bodies, wherewithall they claspe and begirt the body of the Elephant, continually byting of him, untill he fall downe dead, and in the which fall they are also bruzed to peeces; for the safeguard of themselves, they have this device. They get and hide themselves in trees, covering their head, and letting the other part hang downe like a rope: in those trees they watch untill the Elephant come to eate and croppe of the branches; then, suddenly, before he be aware, they leape into his face, and digge out his eyes, then doe they claspe themselves about his necke, and with their tayles, or hinder parts, beate and vexe the Elephant, untill they have made him breathlesse, for they strangle him with theyr fore parts, as they beate them with the hinder, so that in this combat they both perrish: and this is the disposition of the Dragon, that he never setteth upon the Elephant, but with the advantage of the place, and namely from some high tree or Rocke.

"Sometimes againe, a multitude of dragons doe together observe the pathes of the Elephants, and crosse those pathes they tie together their tailes as it were in knots, so that when the Elephant commeth along in them, they insnare his legges, and suddainly leape uppe to his eyes, for that is the part they ayme at above all other, which they speedily pull out, and so not being able to doe him any more harme, the poore beast delivereth himselfe from present death by his owne strength, and yet through his blindnesse received in that combat, hee perrisheth by hunger, because he cannot choose his meate by smelling, but by his eyesight."

The Crocodile.

The largest of the Saurians which we have left us, is the Crocodile; and it formerly had the character of

being very deceitful, and, by its weeping, attracted its victims. Sir John Mandeville thus describes them:—
"In this land, and many other places of Inde, are many cocodrilles, that is a maner of a long serpent, and on nights they dwell on water, and on dayes they dwell on land and rocks, and they eat not in winter. These serpents sley men, and eate them weeping, and they have no tongue."

On the contrary, the Crocodile has a tongue, and a very large one too. As to the fable of its weeping, do we not even to this day call sham mourning, "shedding crocodile's tears?" Spenser, in his "Faerie Queene," thus alludes to its supposed habits (B. I. c. 5. xviii.):—

> "As when a wearie traveller, that strayes
> By muddy shore of broad seven-mouthed Nile,
> Unweeting of the perillous wandring wayes,
> Doth meete a cruell craftie crocodile,
> Which in false griefe hyding his harmeful guile,
> Doth weepe full sore, and sheddeth tender tears.
> The foolish man, that pities all this while
> His mourneful plight, is swallowed up unawares,
> Forgetfull of his owne, that mindes another's cares."

And Shakespeare, from whom we can obtain a quotation on almost anything, makes Othello say (Act iv. sc. 1):—

> "O devil, devil!
> If that the earth could teem with woman's tears,
> Each drop she falls would prove a crocodile;—
> Out of my sight!"

Gesner, and Topsell, in his "Historie of Four-Footed Beastes," give the accompanying illustration of a hippopotamus eating a crocodile, the original of which, they say, came from the Coliseum at Rome, and was then in the Vatican.

Topsell, in his "History of Serpents," dwells lovingly, and lengthily, on the crocodile. He says:—"Some have written that the Crocodile runneth away from a man if he winke with his left eye, and looke steadfastly uppon him with his right eye, but if this bee true, it is not to be attributed to the vertue of the right eye, but onely to the rarenesse of sight, which is conspicuous to the Serpent from one eye. The greatest terrour unto Crocodiles, as both *Seneca* and *Pliny* affirme, are the

inhabitants of the Ile *Tentyrus* within *Nilus*, for those people make them runne away with their voyces, and many times pursue and take them in snares. Of these people speaketh *Solinus* in this manner:—There is a generation of men in the Ile *Tentyrus* within the waters of *Nilus*, which are of a most adverse nature to the Crocodile, dwelling also in the same place. And, although their persons or presence be of small stature, yet heerein is theyr courage admired, because at the suddaine sight of a Crocodile, they are no whit daunted; for one of these dare meete and provoke him to runne away. They will also leape into Rivers and swimme after the Crocodile, and, meeting with it, without feare cast themselves uppon the Beasts backe, ryding on him as uppon a horse. And if the Beast lift uppe his head to byte him, when hee gapeth they put into his mouth a wedge, holding it hard at both ends with both their hands, and so, as it were with a bridle, leade, or rather drive, them captives to the Land, where, with theyr noyse, they so terrifie them, that they make them cast uppe the bodies which they had swallowed into theyr bellies; and because of this antypathy in Nature, the Crocodyles dare not come neare to this Iland.

"And *Strabo* also hath recorded, that at what time crocodiles were brought to Rome, these *Tentyrites* folowed and drove them. For whom there was a certaine great poole or fish-pond assigned, and walled about, except one passage for the Beast to come out of the water into the sun shine : and when the people came to see them, these *Tentyrites*, with nettes would draw them to the Land, and put them backe againe into the water at theyr owne pleasure. For they so hooke them by theyr eyes, and bottome of their bellyes, which are their tenderest partes, that, like as horses broken by theyr Ryders, they yeelde unto them, and forget theyr strength in the presence of these theyr Conquerors. . . .

"To conclude this discourse of Crocodiles inclination, even the Egyptians themselves account a Crocodile a savage, and cruell murthering beast, as may appeare by their Hieroglyphicks, for when they will decypher a mad man, they picture a Crocodile, who beeing put from his desired prey by forcible resistance, hee presently rageth against himselfe. And they are often taught by lamentable experience, what fraude and malice to mankind liveth in these beasts; for, when they cover themselves under willowes and greene hollow bankes, till some people come to the waters side to draw and fetch water, and then suddenly, or ever they be aware, they are taken, and drawne into the water.

"And also, for this purpose, because he knoweth that he is not able to overtake a man in his course or chase, he taketh a great deale of water in his mouth, and casteth it in the pathwaies, so that when they endeavour to run from the crocodile, they fall downe in the slippery path, and are overtaken and destroyed by him. The common proverbe also, *Crocodili lachrimæ*, the Crocodile's teares, justifieth the treacherous nature of this beast, for there are not many bruite beasts that can weepe, but such is the nature of the Crocodile, that to get a man within his danger, he will sob, sigh, and weepe, as though he were in extremitie, but suddenly he destroyeth him. Others say, that the Crocodile weepeth after he hath devoured a man. . . .

"Seeing the friendes of it are so few, the enemies of it must needes be many, and therefore require a more large catalogue or story. In the first ranke whereof commeth (as worthy the first place), the *Ichneumon* or *Pharaoh's Mouse*, who rageth against their egges and their persons; for it is certaine that it hunteth with all sagacity of sense to find out theyr nests, and having found them, it spoyleth, scattereth, breaketh, and emptieth all theyr egs. They also watch the old ones a sleepe, and finding their mouths open against the beames of the Sunne, suddenly enter into them, and, being small, creepe downe theyr vast and large throates before they be aware, and then, putting the Crocodile to exquisite and intollerable torment, by eating their guttes asunder, and so théir soft bellies, while the Crocodile tumbleth to and fro sighing and weeping, now in the depth of water, now on the Land, never resting till strength of nature fayleth. For the incessant gnawing of the *Ichneumon* so provoketh her to seek her rest, in the unrest of every part, herbe, element, throwes, throbs, rowlings, tossings, mournings, but all in vaine, for the enemy within her breatheth through her breath, and sporteth herselfe in the consumption of those vitall parts, which wast and weare away by yeelding to her unpacificable teeth, one after the other, till shee that crept in by stealth at the mouth, like a puny theefe, come out at the belly like a conquerour, thorough a passage opened by her owne labour and industry. . . .

"The medicines arising out of it are also many. The first place belongeth to the Caule, which hath moe benefits or vertues in it, than can be expressed. The bloud of a Crocodile is held profitable for many thinges, and among other, it is thought to cure the bitings of any Serpent. Also by annoynting the eyes, it cureth both the dregs, or spots of blood in them, and also restoreth soundnesse and clearenesse to the sight, taking away all dulnesse, or deadnesse from the eyes. And it is said, that if a man take the liquor which commeth from a piece of a Crocodyle fryed, and annoynte therewithall his wound or harmed part, that then he shall bee presently rid of all paine and torment. The skinne both of the Land and Water Crocodile dryed into powder, and the same powder, with Vineger or Oyle, layd upon a part or member of the body, to be seared, cut off or lanced, taketh away all sense and feeling of paine from the instrument in the action.

"All the Ægytians doe with the fat or sewet of a Crocodile, (*is to*) annoynt all them that be sick of Feavers, for it hath the same operation which the fat of a Sea-dogge, or Dog-fish hath, and, if those parts of men and beasts which are hurt and wounded with Crocodile's teeth, be annoynted with this fat, it also cureth them. Being concocted with Water and Vineger, and so rowled uppe and downe in the mouth, it cureth the tooth-ache: and also it is outwardly applyed agaynst the byting of Flyes, Spyders, Wormes, and such like, for this cause, as also because it is thought to cure Wennes, bunches in the flesh, and olde woundes. It is solde deare, and held pretious in *Alcair*, (Cairo.) *Scaliger* writeth that it cureth the *Gangren*. The Canyne teeth which are hollow, filled with Frankinsence, and tyed to a man or woman, which hath the toothach, cureth them, if the party know not of the carrying them about: And so they write, that if the little stones which are in their belly be taken forth and so used, they work the same effect against Feavers. The dung is profitable against the falling off of the hayre, and many such other things."

THE BASILISK AND COCKATRICE.

Aldrovandus portrays the Basilisk with eight legs. Topsell says it is the same as the Cockatrice, depicts it as a crowned serpent, and says :—" This Beast is called by the Græcian *Baziliscos*, and by the Latine, *Regulus*, because he seemeth to be the King of Serpents, not for his magnitude or greatnesse : For there are many Serpents bigger than he, as there be many foure-footed Beastes bigger than the Lyon, but, because of his stately pace, and magnanimious mind : for hee creepeth not on the earth like other Serpents, but goeth halfe upright, for which occasion all other Serpentes avoyde his sight. And it seemeth nature hath ordayned him for that pur-

pose ; for, besides the strength of his poyson, which is uncurable, he hath a certain combe or Corronet uppon his head, as shall be shewed in due place."

Pliny thus describes " The Serpents called Basilisks. There is the same power [1] also in the serpent called the

[1] Alluding to the Catoblepon (see ante, p. 85), and its power of killing animals and human beings with its eye. This power does not seem confined to animals, for Sir John Mandeville says :—"An other yle there is northward where there are many evill and fell women, and they have precious stones in their eies, and they have such kinde yᵗ if they behold any man with wrath, they sley them of the beholding, as the Basalisk doeth."

Basilisk. It is produced in the province of Cyrene, being not more than twelve fingers in length. It has a white spot on the head, strongly resembling a sort of diadem. When it hisses, all the other serpents fly from it: and it does not advance its body, like the others, by a succession of folds, but moves along upright and erect upon the middle. It destroys all shrubs, not only by its contact, but even those that it has breathed upon; it burns up all the grass too, and breaks the stones, so tremendous is its noxious influence. It was formerly a general belief that if a man on horseback killed one of these animals with a spear, the poison would run up the weapon and kill, not only the rider, but the horse as well. To this dreadful monster the effluvium of the weasel is fatal, a thing which has been tried with success, for kings have often desired to see its body when killed; so true is it that it has pleased Nature that there should be nothing without its antidote. The animal is thrown into the hole of the basilisk, which is easily known from the soil around it being infected. The weasel destroys the basilisk by its odour, but dies itself in this struggle of nature against its own self."

Du Bartas says:—

> "What shield of Ajax could avoid their death
> By th' Basilisk whose pestilentiall breath
> Doth pearce firm Marble, and whose banefull eye
> Wounds with a glance, so that the wounded dye."

The origin of the Cockatrice is, to say the least, peculiar:—"There is some question amongst Writers, about the generation of this Serpent: for some, (and those very many and learned,) affirme him to be brought forth of a Cockes egge. For they say that when a Cocke groweth old, he layeth a certaine egge without any shell, instead whereof it is covered with a very thicke skinne, which is able to withstand the greatest force of an easie blow or fall. They say, moreover, that this Egge is layd onely in the Summer time, about the beginning of the Dogge-dayes, being not so long as a Hens Egge, but round and orbiculer: Sometimes of a Foxie, sometimes of a yellowish muddy colour, which Egge is generated of the putrified seed of the Cocke, and afterward sat upon by a Snake or a Toad, bringeth forth the Cockatrice, being halfe a foot in length, the hinder part like a Snake, the former part like a Cocke, because of a treble combe on his forehead.

"But the vulger opinion of Europe is, that the Egge is nourished by a Toad, and not by a Snake; howbeit, in better experience it is found that the Cocke doth sit on that egge himselfe: whereof *Levinus Lemnius* in his twelfth booke of the hidden miracles of nature, hath this discourse, in the fourth chapter thereof. There happened (saith he) within our memory in the Citty *Pirizæa*, that there were two old Cockes which had layd Egges, but they could not, with clubs and staves drive them from the Egges, untill they were forced to breake the egges in sunder, and strangle the Cockes. . . .

"There be many grave humaine Writers, whose authority is irrefragable, affirming not onely that there be cockatrices, but also that they infect the ayre, and kill with their sight. And *Mercuriall* affirmeth, that when he was with *Maximilian* the Emperour, hee saw the carkase of a cockatrice, reserved in his treasury among his undoubted monuments. . . . Wee doe read that in Rome, in the dayes of Pope *Leo* the fourth (847 to 855), there was a cockatrice found in a Vault of a Church or Chappell, dedicated to Saint *Lucea*, whose pestiferous breath hadde infected the Ayre round about, whereby great mortality followed in Rome: but how the said Cockatrice came thither, it was never knowne. It is most probable that it was created, and sent of GOD for the punnishment of the Citty, which I do the more easily beleeve, because *Segonius* and *Julius Scaliger* do affirme, that the sayd pestiferous beast was killed by the prayers of the said *Leo* the fourth. . . .

"The eyes of the Cockatrice are redde, or somewhat inclyning to blacknesse; the skin and carkase of this beast have beene accounted precious, for wee doe read that the *Pergameni* did buy but certaine peeces of a Cockatrice, and gave for it two pound and a halfe of Sylver: and because there is an opinion that no Byrd, Spyder, or venomous Beast will endure the sight of this Serpent, they did hang uppe the skinne thereof stuffed, in the Temples of *Apollo* and *Diana*, in a certaine thinne Net made of Gold; and therefore it is sayde, that never any Swallow, Spider, or other Serpent durst come within those Temples; And not onely the skinne or the sight of the Cockatrice worketh this effect, but also the flesh thereof, being rubbed uppon the pavement, postes, or Walles of any House. And moreover, if Silver bee rubbed over with the powder of the Cockatrices flesh, it is likewise sayde that it giveth it a tincture like unto Golde: and, besides these qualities, I remember not any other in the flesh or skinne of this serpent. . . .

"We read also that many times in *Affrica*, the Mules fall downe dead for thirst, or else lye dead on the ground for some other causes, unto whose Carkase innumerable troupes of Serpentes gather themselves to feede there uppon; but when the Bazeliske windeth the sayd dead body, he giveth forth his voyce : at the first hearing whereof, all the Serpents hide themselves in the neare adjoyning sandes, or else runne into theyr holes, not daring to come forth againe, untill the Cockatrice have well dyned and satisfied himselfe. At which time he giveth another signall by his voyce of his departure : then come they forth, but never dare meddle with the remnants of the dead beast, but go away to seeke some other prey. And if it happen that any other pestiferous beast cometh unto the waters to drinke neare the place wherein the Cockatrice is lodged, so soone as he perceiveth the presence thereof, although it be not heard nor seene, yet it departeth back againe, without drinking, neglecting his owne nutriment, to save itselfe from further danger : whereupon *Lucanus* saith,

———*Late sibi submovet omne*
Vulgus, et in vacua regnat Basiliscus arena.

Which may be thus englished ;

He makes the vulgar farre from him to stand,
While Cockatrice alone raignes on the sand.

" Now we are to intreate of the poyson of this serpent, for it is a hot and a venemous poyson, infecting the Ayre round about, so as no other Creature can live neare him, for it killeth, not onely by his hissing, and by his sight, (as is sayd of the Gorgons) but also by his touching, both immediately, and mediately ; that is to say, not onely when a man toucheth the body it selfe, but also by touching a Weapon wherewith the body was slayne, or any other dead beast slaine by it, and there is a common fame, that a Horseman taking a Speare in his hand, which had beene thrust through a Cockatrice,

CURIOUS CREATURES.

did not onely draw the poyson of it unto his owne body, and so dyed, but also killed his horse thereby."

THE SALAMANDER.

Many writers have essayed this fabled creature, but almost all have approached the subject with diffidence, as if not quite sure of the absolute entity of the animal. Thus, Aristotle does not speak of it authoritatively:—
"And the Salamander shews that it is possible for some animal substances to exist in the fire, for *they say* that fire is extinguished when this animal walks over it." Pliny, on Salamanders, writes:—"We find it stated by many authors, that a serpent is produced from the

spinal marrow of a man. Many creatures, in fact, among the quadrupeds even, have a secret, and mysterious origin.

"Thus, for instance, the salamander, an animal like a lizard in shape, and with a body starred all over, never comes out except during heavy showers, and disappears the moment it becomes fine. This animal is so intensely cold as to extinguish fire by its contact, in the same way that ice doth. It spits forth a milky matter from its mouth; and whatever part of the human body is touched with this, all the hair falls off, and the part assumes the appearance of leprosy. . . . The wild boar of Pamphylia, and the mountainous parts of Cilicia, after having devoured a Salamander, will become poisonous to those who eat its flesh ; and yet the danger is quite imperceptible by reason of any peculiarity in the smell and taste. The Salamander, too, will poison either water or wine in which it happens to be drowned ; and, what is more, if it has only drunk thereof, the liquid becomes poisonous."

This idea of an animal supporting life in the fire is not confined to the Salamander alone, for both Aristotle and Pliny aver that there is a fly which possesses this accomplishment. Says the former :—" In Cyprus, when the manufacturers of the stone called *chalcitis* burn it for many days in the fire, a winged creature something larger than a great fly is seen walking and leaping in the fire : these creatures perish when taken from the fire." And the latter :—" That element, also, which is so destructive to matter, produces certain animals ; for in the copper-smelting furnaces of Cyprus, in the very midst of the fire, there is to be seen, flying about, a four-footed animal with wings, the size of a large fly : this creature, called the 'pyrallis,' and by some the 'pyrausta.' So long as it remains in the fire it will live, but if it comes out, and flies a little distance from it, it will instantly die."

Ser Marco Polo thoroughly pooh-poohs the idea of the Salamander, and says it is Asbestos. Speaking of the Province of Chingintalas, he says :—" And you must know that in the same mountain there is a vein of the substance of which Salamander is made. For the real truth is that the Salamander is no beast, as they allege in our part of the world, but is a substance found in the earth ; and I will tell you about it.

" Everybody must be aware that it can be no animal's nature to live in fire, seeing that every animal is composed of all the four elements. Now, I, Marco Polo, had a Turkish acquaintance of the name of Zurficar, and he was a very clever fellow, and this Turk related to Messer Marco Polo how he had lived three years in that region on behalf of the Great Kaan, in order to procure those Salamanders for him. He said that the way they got them was by digging in that mountain till they found a certain vein. The substance of this vein was then taken and crushed, and, when so treated, it divides, as it were, into fibres of wool, which they set forth to dry. When dry, these fibres were pounded in a great copper mortar, and then washed, so as to remove all the earth, and to leave only the fibres, like fibres of wool. These were then spun, and made into napkins. When first made, these napkins are not very white, but by putting them in the fire for a while they come out as white as snow. And so again, whenever they become dirty they are bleached by being put in the fire.

"Now this, and nought else, is the truth about the Salamander, and the people of the country all say the same. Any other account of the matter is fabulous nonsense. And I may add that they have, at Rome, a napkin out of this stuff, which the Grand Kaan sent to the Pope, to make a wrapper, for the Holy Sudarium of Jesus Christ."

That extremely truthful person, Benvenuto Cellini, in his thoroughly veracious autobiography, tells us the following *Snake Story:*—" When I was about five years old, my father happened to be in a basement-chamber of our house, where they had been washing, and where a good fire of oak-logs was still burning; he had a viol in his hand, and was playing and singing alone beside the fire.

"The weather was very cold. Happening to look into the fire, he spied in the middle of those most burning flames a little creature like a lizard, which was sporting in the core of the intensest coals. Becoming instantly aware of what the thing was, he had my sister and me called, and, pointing it out to us children, gave me a great box on the ears, which caused me to howl and weep with all my might. Then he pacified me good-humouredly, and spoke as follows: 'My dear little boy, I am not striking you for any wrong that you have done, but only to make you remember that that lizard which you see in the fire is a salamander, a creature which has never been seen before, by any one of whom we have credible information.' So saying, he kissed me, and gave me some pieces of money."

Even Topsell is half-hearted about its fire-resisting qualities, giving no modern instances, and only, for it, quoting old authors. According to his account, and to the picture which I have taken from him, the Salamander is not a prepossessing-looking animal:—"The Salamander is also foure-footed like a Lyzard, and all the body over it is set with spots of blacke and yellow, yet is the sight of it abhominable, and fearefull to man. The head of it is great, and sometimes they have yellowish bellyes and tayles, and sometimes earthy."

He also says its bite is not only poisonous, but incurable, and that it poisons all it touches.

The Toad.

Toads were always considered venomous and spiteful, and they had but one redeeming quality, which seems to be lost to its modern descendants:—

> "Sweet are the uses of adversity;
> Which, like the toad, ugly and venomous,
> Wears yet a precious jewel in his head."
> (*As You Like It*, Act ii. sc. 1.)

Pliny says of these animals:—" Authors quite vie with one another in relating marvellous stories about them; such, for instance, as that if they are brought into the midst of a concourse of people, silence will instantly prevail; as also that, by throwing into boiling water, a small bone that is found in their right side, the vessel will immediately cool, and the water refuse to boil again until it has been removed. This bone, they say, may be found by exposing a dead toad to ants, and letting them eat away the flesh; after which the bones must be put into the vessel one by one.

"On the other hand, again, in the left side of this reptile there is another bone, they say, which, when thrown into water, has all the appearance of making it boil, and the name given to which is 'apocynon' (*averting dogs*). This bone it is said has the property of assuaging the fury of dogs, and, if put in the drink, of conciliating love, and ending discord and strife. Worn, too, as an amulet, it acts as an aphrodisiac, we are told."

Topsell writes so diffusely on the virtues of these "toad stones" that I can only afford space for a portion of his remarks:—"There be many late Writers, which doe affirme that there is a precious stone in the head of a Toade, whose opinions (because they attribute much to the vertue of this stone) is good to examine in this place. . . . There be many that weare these stones in Ringes, beeing verily perswaded that they keepe them from all manner of grypings and paines of the belly, and the small guttes. But the Art, (as they term it) is in taking of it out, for they say it must be taken out of the head alive, before the Toade be dead, with a peece of cloth of the colour of redde Skarlet, wherewithall they are much delighted, so that while they stretch out themselves as it were in sport upon that cloth, they cast out the stone of their head, but instantly they sup it up againe, unlesse it be taken from them through some secrete hole in the said cloth, whereby it falleth into a cesterne or vessell of water, into the which the Toade dare not enter, by reason of the coldnes of the water. . . .

"This stone is that which in aunciente time was called *Batrachites*, and they attribute unto it a vertue besides the former, namely, for the breaking of the stone in the bladder, and against the Falling sicknes. And they further write that it is a discoverer of present poyson, for in the presence of poyson it will change the colour. And this is the substaunce of that which is written about this stone. Now for my part I dare not conclude either with it, or against it, for many are directlie for this stone ingendered in the braine or head of the Toade: on the other side, some confesse such a stone by name and nature, but they make doubt of the generation of it, as others have delivered; and therefore, they beeing in sundry opinions, the hearing whereof might confound the Reader, I will referre him for his satisfaction unto a Toade, which hee may easily every day kill: For although when the Toade is dead, the vertue thereof be lost, which consisted in the eye, or blew spot in the middle, yet the substance remaineth, and, if the stone be found there in substance, then is the question at an end; but, if it be not, then must the generation of it be sought for in some other place."

CURIOUS CREATURES.

The Leech.

The Leech has, from a very early age, been used as a means of letting blood; but, among the old Romans, it had medicinal uses such as we know not of now. It was used as a hair dye. Pliny gives two receipts for making it, and it must have been powerful stuff, if we can believe his authority :—" Leeches left to putrify for forty days in red wine, stain the hair black. Others, again, recommend one sextarius of leeches to be left to putrefy the same number of days in a leaden vessel, with two sextarii of vinegar, the hair to be well rubbed with the mixture in the sun. According to Sornatius this preparation is, naturally, so penetrating, that if females, when they apply it, do not take the precaution of keeping some oil in the mouth, the teeth, even, will become blackened thereby."

Olaus Magnus gives us the accompanying picture of the

luxurious man in his arm-chair by the river-side, catching his own leeches, and suffering from gnats; and also his far more prudent friend, who makes the experiment on the vile body of his horse, and thus saves his own blood; but he gives us no account of its habits and customs.

The Scorpion.

Of the Scorpion, Pliny says :—" This animal is a dangerous scourge, and has a venom like that of the serpent; with the exception that its effects are far more painful, as the person who is stung will linger for three days before death ensues. The sting is invariably fatal to virgins, and nearly always so to matrons. It is so to men also, in the morning, when the animal has issued from its hole in a fasting state, and has not yet happened to discharge its poison by an accidental stroke. The tail is always ready to strike, and ceases not for an instant to menace, so that no opportunity may possibly be lost. . . .

"In Scythia, the Scorpion is able to kill even the swine, with its sting, an animal which, in general, is proof against poisons of this kind in a remarkable degree. When stung, those swine which are black, die more speedily than others, and more particularly if they happen to throw themselves into the water. When a person has been stung, it is generally supposed that he may be cured by drinking the ashes of the Scorpion mixed with wine. It is the belief also that nothing is more baneful to the Scorpion than to dip it in oil. . . . Some writers, too, are of opinion that the Scorpion devours its offspring, and that the one among the young which is most adroit avails itself of its sole mode of escape, by placing itself on the back of the mother, and thus finding a place where it is in safety from the tail and sting. The one that thus escapes, they say, becomes the avenger of the rest, and, at last, taking advantage of its elevated position, puts its parents to death."

Topsell has some marvels to relate concerning the generation of Scorpions :—" And it is reported by *Elianus*, that about *Estamenus* in India, there are abundance of Scorpions generated, onely by corrupt raine water standing in that place. Also, out of the Baziliske beaten into peeces, and so putrified, are Scorpions engendred. And when as one had planted the herbe *Basilica* on a wall, in the roome or place thereof hee found two Scorpions. And some say that if a man chaw in his mouth, fasting, this herbe Basill before he wash, and, afterwards, lay the same abroade uncovered where no sun commeth at it for the space of seaven nights, taking it in all the daytime, he shall at length find it transmuted into a Scorpion, with a tayle of seaven knots.

"*Hollerius*, to take away all scruple of this thing, writeth that in Italy, in his dayes, there was a man that had a Scorpion bredde in his braine, by continuall smelling to this herbe Basil; and *Gesner* by relation of an Apothecary in Fraunce, writeth also a storie of a young mayde, who by smelling to Basill, fell into an exceeding head-ach, whereof she died without cure, and, after her death, beeing opened, there were found little Scorpions in her braine.

"*Aristotle* remembreth an herbe which he calleth *Sisimbriæ*, out of which putrified Scorpions are engendered. And wee have showed already, in the history of the Crocodile, that out of the Crocodile's egges doe many times come Scorpions, which at their first egression doe kill theyr dam that hatched them."

There is a curious legend, that if a Scorpion is surrounded by fire, so that it cannot escape, it will commit suicide by stinging itself to death.

The Ant.

No one would credit the industrious Ant, whose ways we are told to consider, and gather wisdom therefrom, was avaricious and lustful after gold ; but it seems it was even so, at least, in Pliny's time ; but then they were abnormally large :—" The horns of an Indian Ant, suspended in the temple of Hercules at Erythræ (*Ritri*) have been looked upon as quite miraculous for their size. This ant excavates gold from holes, in a country to the north of India, the inhabitants of which are known as the Dardæ. It has the colour of a cat, and is in size as large as an Egyptian wolf. This gold, which it extracts in the winter, is taken by the Indians during the heats of summer, while the Ants are compelled, by the excessive warmth, to hide themselves in their holes. Still, however, on being aroused by catching the scent of the Indians, they sally forth, and frequently tear them to pieces, though provided with the swiftest Camels for the purpose of flight ; so great is their fleetness, combined with their ferocity, and their passion for gold ! "

The Bee.

The Busy Bee, too, according to Olaus Magnus, developed, in the regions of the North, a peculiarity to which it seems a stranger with us, but which might be encouraged, with beneficial effect, by the Temperance Societies.

The Bees infested drunkards, being drawn to them

by the smell of the liquor with which they had soaked their bodies, and stung them.

THE HORNET.

So also, up North, they seem to have had a special

breed of Hornets, which must have been ferocious indeed, sparing neither man nor beast, as is evidenced by the corpses, and by the extremely energetic efforts of the yet living man to cope with his enemies.

INDEX.

INDEX.

ABAMIRON, *country of men with legs reversed*, 9.
Acanthis, the, 70.
Accursius, 147.
Achillium. See *Sponges*.
Ædonaus, 287.
Ægipanæ, *a name for Satyrs*, 57.
Ægithus, the, 70, 71.
Ægopithecus, the, 55.
Ælianus, 88, 93, 96, 148, 158, 212, 280, 331.
Æsalon, the, 70.
Æsculapius, 148.
Ætolia, 280.
Agatharcides, 10, 16.
Aïnos, the, *a hairy people of Japan*, 50, 51.
Albertus, 93, 100, 252.
Albinos, 10.
Alciatus, 65.
Aldrovandus, 47, 48, 81, 97, 154, 170, 171, 172, 179, 180, 204, 228, 256, 262, 270, 302, 317.
Alexander, 146.
Alumnus, 100.
Amahut, *a tree*, 67.
Amazons, 23, *their fate after their defeat by the Greeks*, 24, 25. *Sir John Mandeville's account of them*, 25, 26; *called Medusæ*, 85.

Ambergris, 222, 223.
Anclorus, the, 148.
Andrew, *an Italian*, 151.
Androgyni, *tribe of*, 11.
Animal lore, 67, 68, 69, 70, 71.
Ant, the, 71, 112, 332.
Antacæi (*whales without spinal bones*), 226.
Antelope, the, 145, 146.
Anthropophagi, 6, 9, 10, 18, 72.
Anthus, the, 71.
Anu, 80.
Apes, 65, 66.
Apocynon. See *The Toad*.
Apollonides, 12.
Apollonius, 58, 59.
Archelaüs, 21.
Archigene, 134.
Arctopithecus, the, *or Bear-Ape*, 55, 66.
Arimaspi, 8, 9.
Aristotle, 71, 105, 148, 156, 199, 201, 203, 248, 253, 262, 268, 286, 287, 323, 324, 331.
Artemidorus, 16.
Asbestos. See *Salamander*.
Astomi, *a people with no mouths, and who subsist by smell*, 15.
Ass, the, 70.
Ass, the Indian, 88.
Ass, the wild, 68.
Atergatis, 209.

INDEX.

Athenæus, 86.
Ausonius, 64.
Avicen, 72, 287.

B.

BABOONS, 62.
Bacchantes, 80.
Bacchæ, *a name for Satyrs*, 56.
Baffin, 245.
Balæna, the, 239, 240.
Barnacle Goose, the, 174, 175, 176, 177, 178, 179.
Bartlemew de Glanville, 231.
Basilisk, 156, 317, 318, 319, 321, 331.
Batrachites. See *The Toad*.
Bear, the, 68, 104, 105, 106, 107, 108, 109, 110, 111, 112, 113, 114, 115, 116, 117, 118, 119, 120, 121, 122, 123, 124, 125, 148.
Bear-Ape. See *Arctopithecus*.
Bee, the, 112, 113, 332, 333.
Beeton, 10.
Bekenhawh, 189.
Bellonius, Petrus, 96.
Berosus, 79, 206.
Bevis of Hampton, 158.
Bird, Miss, 50.
Birds, peculiarities of, 204, 206.
Bishop-fish, the, 228, 230.
Boar, the wild, 69, 111, 139.
Boas, the, 289, 290, 291.
Bolindinata. See *Bird of Paradise*.
Boloma, the. See *Dog-fish*.
Bonosa, the, 193.
Bœothius, 228.
Borometz, the. See *Lamb Tree*.
Boscawen, W. St. Chad, 78.
Brazavolus, 94.
Bugil, the, 84.
Bull, the, *and Bears*, 109, *and Wolves*, 137.
Bustard, the, 148.

C.

CADAMUSTUS, ALOISIUS, 278.
Cadmus, 64, 65.
Cæsar, Julius, 46, 47, 148.
Calf and Wolves, 137.
Calingæ, *a tribe of India whose women conceive at the age of five years and die at eight*, 17.
Callimachus, 285.
Calliphanes, 11.
Cambden, Mr., 144.
Camden, 177.
Camel, the, 148.
Canis Lucernarius, 150, 151.
Cardanus, Hieronimus, 53, 226, 287, 291, 305.
Cartazonon. See *Unicorn*.
Carthier, Jacques, 237.
Cat, the, 154, 155, 156.
Caterpillar, the, 71.
Catharcludi, *a tribe in India*, 14.
Catableponta, *name for Gorgon*, 84, 85, 318.
Cattle, *curious*, 23.
Cebi, the, 57.
Cellini, Benvenuto, 325, 326.
Centaurs, 65, 78, 79, 80, 81, 82, 83.
Cephus, the, 74.
Cercopithecus, the, 52, 53.
Cetum Capillatum vel Crinitum. See *Whale, Hairy*.
Chameleon, the, 163.
Chimæra, the, 64, 170, 171.
Chiron, *the Centaur*, 79.
Chloræus, the, 69.
Choromandæ, *a nation without a proper voice*, 15.
Christie, Mr., on *Palæolithic remains*, 39.
Cicero, 12.
Circhos, the, 247.
Claudius, Emperor. See *Orca*.
Clayks. See *Barnacle Geese*.
Clement, Pope, 96.

INDEX.

Clitarchus, 16.
Cock, the, 156, 157.
Cock with serpent's tail, 204, 205.
Cockatrice, the, 85, 317, 319, 320, 321, 322.
Cœlius, 77.
Condor, the, 183.
Conger Eel, the, 262.
Corocotta, the, 72.
Couret, M. de, 5.
Crab, the, 129, 267, 268.
Crane, the, 203.
Crannoges, 41.
Crates of Pergamus, 10, 17.
Crawford, John, 49.
Cray-fish, 267.
Cristotinius. See *Lamia*.
Crocodile, the, 311, 312, 313, 314, 315, 316, 317.
Crocotta, the, 159.
Cronos, or Hea, 79.
Crow, the, 70, 129, 130, 131.
Ctesias, 4, 14, 16, 71.
Cuvier, 185.
Cyclops, 7, 65.
Cynocephalus, the, 55, 56, 63.
Cyrni, the, *who live 400 years*, 15.

D.

DÆDALUS, H.M.S., 274, 275, 276.
Dagon, 209.
Damon, 12.
Darwin, *Descent of Man*, 1; Tailed men, 4; *Shell-fish middens in Tierra del Fuego*, 42.
Davis, Barnard, 50.
De Barri, Gerald, 174.
Deer and Bears, 109.
De Leo, Ronzo, 31.
Demetrius, 121, 237.
Democritus, 131, 285, 306.

Denbigh Worme, the. See *Dragons*.
Descent of Man, 1.
De Thaun, Philip, 91.
De Veer, Gerat, 177.
Devil Whale, the. See *Trol Whale*.
Dingo, the, 126.
Dinornis Giganteus. See *Moa*.
Dion, 77.
Dog, the, 150, 151, 152, 153, 154.
Dog-fish, the, 255.
Dog, *the Mimic or Getulian*, 150, 151.
Dolphin, the, 242, 243.
Dordogne, *Palæolithic remains in caves at*, 39.
Dormouse, the, 67.
Draco, 64.
Dracontopides. See *Dragons*.
Dragon, the, 158, 162, 293, 294, 295, 296, 297, 298, 299, 300, 301, 302, 303, 304, 305, 306, 307, 308, 309, 310, 311.
Drake, Sir Francis, 177.
Du Bartas, 74, 168, 169, 179, 185, 186, 200, 202, 225, 230, 231, 243, 319.
Duck, the, 70; *four-footed*, 203.
Dugong, the, 213.
Duret, Claude, 166.
Dwarfs, *with no mouth*, 19; *mentioned in the Bible*, 26; *Homer and the pygmies—battle with the Cranes*, 26, 27, 28; *only twenty-seven inches high*, 28; *their age*, 28; *Spurious pygmies*, 28; *Northern dwarfs*, 29; *in America*, 29, 30, 31; *African dwarfs*, 31, 32; *their acuteness*, 33.

E.

EAGLE, the, 69, 70.

INDEX.

Eale, the, 159, 160.
Echeneis, the. See *Remora*.
Edmund, St., 139, 140.
Eels, *thirty feet long*, 18.
Egede, Hans, 270.
Egemon, 280.
Egg, Remarkable, 179, 180.
Ehannum. See *Lamia*.
Eigi - einhamir. See *Were Wolves*.
Elephant, the, 100, 147, 163, 310, 311.
Elpis, 158.
Embarus, 123.
Emin Pacha, 32.
Empusæ. See *Lamia*.
Enchanters, *families of*, 11.
Epyornis maximus, 183.
Ethiopia, *wonders of*, 13.
Eudoxus, 15.
Euryale, 85.

F.

FABRICIUS, GEORGE, 61.
Falisci, or Hirpi, *a tribe unharmed by fire*, 12.
Farnesius, 90.
Fauns, 5, 56, 57, 60.
Ferrerius, Joannes, 95.
Fincelius, 146.
Fish, curious, 248, 249, 250, 251, 252, 253.
Fish, senses of, 258, 259.
Flavianus, 243.
Florentinus, 287.
Footless birds. See *Apodes*.
Formicæ Lions, 58.
Fox, the, 68, 70, 125, 126, 127, 128, 129, 130, 131, 132, 133, 134.
Fridlevus, 293, 294.
Frobisher, Sir Martin, 245.
Frog, the, 68.
Frotho, 293.

G.

GAEKWAR OF BARODA, 129.
Gambarus, the, 244.
Gazelle, the, 67.
Geese, two-headed wild, 203.
Gellius, or Gyllius, Aulus, 158, 281, 302.
Geryon, 64.
Geskleithron, *dwelling of one-eyed men*, 8.
Gesner, 52, 97, 127, 179, 203, 212, 217, 226, 228, 229, 231, 233, 236, 244, 256, 262, 269, 305, 306, 312, 331.
Getulian Dog, the, 150, 151.
Giants, 13, 16, 17, 32; *their stupidity*, 33; *their sobriety*, 33; *Starchaterus Thavestus*, 33, 34, 35, 36; *Giants mentioned in the Bible*, 36; *height of Adam, &c.*, 37; *Gabbaras*, 37; *Posio and Secundilla*, 37; *Sir John Mandeville's giants*, 37, 38.
Gibson, Edmund, 177.
Giraldus Cambrensis, 77, 174, 175.
Gisbertus Germanus, 227, 228.
Gizdhubar, 78, 79, 80.
Glutton, the. See *Gulo*.
Goat, the, 128, 136.
Goblerus, Justinus, 306.
Gorgon, the, 83, 84, 85, 86, 87.
Gorgon blepen, *sharp-sighted persons*, 86.
Gould, Rev. S. Baring, 141.
Grevinus, 302.
Griffins, 8, 180, 181, 182, 183.
Gryphons, 8, 9, 181.
Guenon, the. See *Haut*.
Guillim, 89, 189.
Gulielmus Musicus, 305.
Gulo, the, 101, 102, 103, 104, 105.
Guy, Earl of Warwick, 157.

INDEX.

Gymnetæ, *who live a hundred years*, 16.

H.

HAAFISCH, the. See *Dog-fish*.
Haarwal, the. See *Whale, Hairy*.
Hakluyt, 237, 245.
Halcyon, the, 199, 200.
Hanno, 86.
Harald, King, 307, 308.
Hare, the, 68, 128.
Harmona, 64.
Harpe, the, *a falcon*, 70.
Harpy, the, 171, 172.
Hauser, Caspar, *a wild man*, 45.
Haut or Hauti, the, 66, 67.
Hawkins, Thos., 301, 302.
Hea, 79, 206, 207, 208, 209.
Hea-bani, 79, 80.
Hedgehog, the, 69, 111, 128.
Hegesidemus, 243.
Helcus, the. See *Sea Calf*.
Helen, 286.
Helladice, 208.
Hens, Woolly, 202.
Hentzner, Paul, 93.
Hermias, 243.
Herodotus, 8, 21, 23, 39, 140, 160, 226.
Heron, the, 70.
Hesiodus, 85.
Hippocentaur, the, 59.
Hippopotamus, the, 161, 312.
Hirpi, or Falisci, *a tribe unharmed by fire*, 12.
Hollerius, 331.
Homer, 75.
Hoopoe, the, 196.
Hornet, the 333, 334.
Horse, the, 112, 138, 146, 147, 148, 149, 150.
Horstius, 227.
Hyæna, the, 74, 132.

Hydra, 64, 291, 292.
Hydrophobia, 152, 153.

I.

IBIS, the, 161.
Ichneumon, the, 70, 202, 315, 316.
Ichthyo Centaurus, the, 212.
Ierom, Saint, 59.
Illyrii, *a tribe having fascination in their eyes*, 12.
Incubi, 60.
India, *Wonders of*, 13.
Isodorus, 100.
Isogonus of Nicæa, 10, 11, 12, 15.
Istar, 80.

J.

JAMES IV. and VI. of Scotland, 88.
Jeduah, the. See *Lamb Tree*.
Jerff. See *Gulo*.
Jocasta, 65.
Jochanan, Rabbi, 166.
Johnöen, Lars, 273.
Jovius, Paulus, 237.
Juba, 21.
Jugurtha, 86.

K.

KHUMBABA, 79.
King-fisher. See *Halcyon*.
Kite, the, 69.
Kjökkenmöddings, 41, 42, 43, 44.
Kraken, the, 244, 261, 262, 263, 264, 265, 266, 292.

L.

LACUS INSANUS, 23.
Laius, 65.
Lake dwellings, 39, 40, 41.

INDEX.

La Madelaine, *Palæolithic remains at*, 39.
Lamb tree, the, 165, 166, 167, 168, 169, 170.
Lambri, *Kingdom of*, 5.
Lambton Worme, the. See *Dragons*.
Lamia, the, 74, 75, 76, 77, 78.
Lane, Mr., 218.
Langa, the, 225.
Lapithæ, 80.
Lapwing, the, 196, 197.
Lee, Henry, 165, 292.
Leech, the, 329, 330.
Lemnius, Levinus, 320.
Lenormant, M., 208.
Leone, Giovanni, 198, 201.
Leonine Monster, a, 227.
Leontophonus, the, 158.
Leontopithecus, the, 55.
Leopard, the, 138.
Leucrocotta, the (see also *Manticora*), 159, 160.
Leviathan, 218.
Licetus, 173, 179.
Licosthenes, 81, 146, 180.
Lilith. See *Lamia*.
Linton Worme, the. See *Dragons*.
Lion, the, 71, 88, 156, 157, 158, 159.
Livingstone, Dr., 31.
Livy, 9.
Lizards, flying, 302.
Lotophagi, *Cattle of*, 160.
Loup-garou. See Were Wolf.
Lucanus, 322.
Lucretius, 157.
Lycanthropy. See Were Wolf.
Lycaon. See Were Wolf.
Lynx, the, 129, 159.

M.

MACHLYÆ, *the tribe of, are androgynous*, 11.

Maclean, Rev. —, 271.
Macrobii, *people who live four hundred years*, 15, 16.
M'Quhæ, Capt., 274, 275, 276.
Magalhaen, 190.
Magnus, Olaus, 29, 33, 104, 108, 127, 141, 176, 182, 187, 188, 194, 214, 219, 221, 223, 227, 231, 232, 233, 236, 237, 241, 244, 245, 251, 255, 256, 260, 262, 264, 266, 269, 285, 293, 329, 332.
Manatee, 213.
Mandeville, Sir John, 17, 21, 25, 28, 37, 169, 175, 181, 202, 249, 312, 318.
Mandi, *who live on locusts*, 16.
Mandragora, 112.
Man-fish, 212, 213, 231.
Mani. See *Sponges*.
Manilius, Senator, 184.
Manticora, the, 71, 72, 73, 74, 159.
Maphoon, *a hairy woman*, 49, 50.
Mappa Mundi, 7, 17.
Marcellinus, 134.
Marcellus, 131, 133, 134, 140, 144, 174.
Marco Polo, 5, 28, 100, 182, 249, 324, 325.
Maricomorion, the. See *Manticora*.
Marion, the. See *Manticora*.
Marius, 86.
Marsi, *the tribe of*, 11.
Martlet, the, 189, 190.
Mechovita, 102, 237.
Megasthenes, 14, 15, 16.
Meir, Rabbi, 167.
Men, *tailed*, 4, 5, 17; *one-eyed*, 8, 18; *with legs reversed*, 9; *with sea-green eyes*, 10, 15; *with white hair*, 10, 14, 16; *eat every other day*, 10;

INDEX. 343

those whose touch cures the sting of serpents, 10; *saliva cures ditto,* 10; *testing the fidelity of wives by means of serpents,* 11; *possessing both sexes,* 11; *families of enchanters,* 11; *with the power of fascination in their eyes,* 12; *with two pupils in each eye,* 12; *whose bodies will not sink in water,* 12; *whose perspiration causes consumption,* 12; *the glance of women with double pupils in their eyes is noxious,* 12; *Indians never expectorate, and are subject to no pains,* 13; *Men eight feet high,* 13, 16; *with feet turned backwards, and eight toes,* 14; *with heads of dogs,* 14; *Women only pregnant once in their lives,* 14, 16; *Men with one leg,* 14, 20; *whose feet shade them from the sun,* 14, 20; *without necks, and eyes in their shoulders,* 14, 19; *large and small feet,* 15; *with holes in their faces instead of nostrils, and flexible feet,* 15; *with no mouths, who subsist by smell,* 15; *who live 400 years,* 15; *living on vipers,* 16; *with no shadow,* 16; *live to 130 years and never seem to get old,* 16; *who live 200 years,* 16; *do not live over 40 years,* 16; *who live on locusts,* 16; *Women bear children at seven years of age,* 16; *Women conceive at five years of age and die in their eighth year,* 17; *Men with ears which cover their bodies,* 17; *twelve feet high,* 17; *live on baboon's milk,* 17; *green and yellow,* 18: *Men eating each other,* 18; *without eyes or nose,* 19; *with mouths in their shoulders,* 19; *cover their faces with their lips,* 19; *Dwarfs with no mouth,* 19; *with ears to their shoulders,* 19; *with horses' feet,* 19; *go on all fours,* 19; *go on their knees,* 19; *live by the smell of wild apples,* 19; *covered with feathers,* 20; *Elephant-headed men,* 20; *feed on serpents and lizards,* 21; *Amazons,* 23, 24, 25, 26; *Pygmies,* 26; *their height,* 28; *Early men,* 38; *their skulls,* 38; *the Stone Age,* 38; *Bronze and Iron Ages,* 39; *Palæolithic remains in caves,* 39; *the Lake men,* 39; *early mention of them,* 39; *their food,* 41; *Kitchen middens,* 41; *their wide range,* 41; *Shell-fish middens in Tierra del Fuego,* 42, 43; *Danish middens,* 44; *Wild men,* 41; *Ancient Britons,* 46, 47; *Hairy men,* 47, 49, 50, 51; *Julia Pastrana,* 47; *Puella pilosa of Aldrovandus,* 47, 48; *Hairy people at Ava,* 49, 50; *the Ainos of Japan,* 50, 51; *Moon Woman,* 180.

Menippus, 74, 75, 76, 152.
Menismini, *who live on baboon's milk,* 17.
Mentor, 158.
Mercuriall, 320.
Mermen and Mermaids, 209, 210, 211, 212, 213, 214.
Meryx, the, 253.
Midas, 58.
Milo, Titus Annius, 251.
Milroy, General, 30.
Milton, 8, 218.
Mimick Dog, the, 150, 151.

Mirage, 17.
Moa, the, 181, 183.
Mole, the, 68.
Monboddo, Lord, 5.
Monk-fish, the, 228, 229.
Monoceros. See *Unicorn*, also *Narwhal*.
Monocoli, *people having but one leg*, 14.
Monster, a, 173.
Moon Woman, 180.
Mormolicæ. See *Lamia*.
Morse, the. See *Walrus*.
Moses Chusensis, 166.
Mucianus, 253.
Münster, Sebastian, 177.
Murex, the, 253, 254.
Musculus, the, 226.
Myrepsus, 132, 134.

N.

NARWHAL, the, 244, 245.
Nasomenes, *the tribe of*, 11.
Nebuchadnezzar, 78.
Nemæan Lion, 64.
Nereids, 210.
Niam Niams, 5.
Nicander, 302.
Nisus, the, 70.
Nymphæ, *a name for Satyrs*, 57.
Nymphodorus, 11.

O.

OANNES, *or Hea*, 206, 207, 208, 209.
Obadja, Rabbi, 167.
Octopus. See *Kraken*.
Odoricus, Friar, 170, 175.
Œdipus, 64, 65.
Olaus Magnus. See *Magnus, O.*
Onisecritus, 16.
Onocentaur, the, 56, 83.
Ophiogenes, 10.

Oppianus, 99, 119.
Orca, the, 239, 240, 241.
Osborne, the Royal Yacht, 276, 277.
Ostridge or Estridge, 148, 197, 198.
Ouran Outan, the, 51, 52.
Ourani Outanis, 4.
Ovid, 140.
Owl, the, 70.
Oxen and Wolves, 137, 138.

P.

PAN, the, *a satyr*, 55, 57.
Pan, the Sea, 212.
Pandore, *live two hundred years*, 16.
Panther, the, 162.
Paradise, Birds of, 190, 191.
Parkinson, John, 168.
Pastrana, Julia, *a hairy woman*, 47.
Pausanias, 65.
Pelican, the, 200, 201.
Pegasus, the, 159.
Pergannes, 16.
Peter, the wild boy, 45.
Peter Martyr, 4.
Petronius, 140.
Phalangium, the, 68, 70, 161.
Pharnaces, *a tribe whose perspiration causes consumption*, 12.
Philostratus, 58.
Phœnix, the, 183, 184, 185, 186.
Pholus, *the Centaur*, 80.
Phylarcus, 12.
Physeter, the, 215, 216, 217.
Pierius, 302.
Pitan, a *tribe living on the smell of wild apples*, 19.
Pithocaris, 139.
Plato, 194.
Plesiosaurus, the, 300, 301.

INDEX.

Pliny, 5, 7, 8, 9, 17, 21, 22, 23, 26, 27, 53. 57, 67, 72, 81, 86, 87, 88, 105, 124, 127, 131, 133, 140, 148, 158, 161, 183, 193, 198, 199, 204, 210, 239, 242, 251, 253, 256, 264, 267, 285, 286, 287, 288, 306, 313, 318, 324, 327, 329, 330, 332.
Plutarch, 151, 281.
Polydamna, 286.
Polypus, the. See *Kraken*.
Poæius, Paulus, 95.
Pomponius, Mela, 140.
Pontoppidan, Erik, 261, 270.
Ponzettus, 154.
Pope, Alex., 26.
Postdenius, 282.
Prister, the, 215, 220.
Psylli, *a race whose saliva cures the sting of serpents*, 10.
Pterodactyl, the, 302.
Ptolemy, 5.
Ptolemy, King, 151.
Purchas, *his Pilgrimage*, 29, 177.
Pygmies. See *Dwarfs*.
Pygmæogeranomachia, *a poem on the battle between the Pygmies and the Cranes*, 26.
Pyrallis, the, 70. See also *Salamander*.
Pyrausta. See *Salamander*.
Pyrrhus, King. *His right great toe cured diseases of the spleen*, 13.

R.

RABBIT, the, 68.
Rasis, 156.
Raven, the, 69, 70, 163.
Ravenna, *Monster at*, 173, 174.
Ravisius, Textor, 180.
Ray, the, 255.

Rayn, the, 197.
Regnerus, 294, 295.
Reineke Fuchs, 126.
Remora, the, 253, 254.
Rhinoceros, 89, 97, 98, 99, 100.
Robinson, Phil, 129.
Rodocanakis, 188, 189.
Rondeletius, 227.
Rosmarus, the. See *Walrus*.
Rossamaka, the. See *Gulo*.
Ruc, Rukh, or Rok. See *Griffin*.

S.

SAHAB, the, 247.
St. John, Mr., 5.
Salamander, 323, 324, 325, 326.
Salusbury, John, 300.
Sargon, 209, 268.
Satyr, the, 14.
Satyr, *the classical*, 53, 56, 57, 58, 59, 60.
Satyrs, 55, 56, 61, 62.
Saw Fish, the, 239.
Saxo, 33, 34, 177.
Scaliger, 131, 317, 321.
Scarus, the, 253.
Schilt-bergerus, 284.
Sciapodæ, *men whose feet shade them from the sun*, 14.
Scirti, *a name for Satyrs*, 57.
Scorpion, the, 69, 330, 331, 332.
Scott, Sir Walter, 270, 271.
Scyritæ, *a tribe in India with holes in their faces instead of nostrils, and flexible feet*, 15.
Sea Animals, various, 231.
Sea Calves, 116, 232, 233.
Sea-Cow, the, 232.
Sea Demon, 212.
Sea Dragon, the, 256.
Sea Hare, 132, 234.
Sea-Horse, the, 233, 234.

INDEX.

Seamew, the, 70.
Sea-Mouse, the, 234.
Sea-Nettle, the, 259, 260.
Sea-Pig, the, 235.
Sea Rhinoceros, the. See *Narwhal*.
Sea Satyr, 212.
Sea Serpent, the, 268, 269, 270, 271, 272, 273, 274, 275, 276, 277.
Sea Unicorn, the. See *Narwhal*.
Seal, the. See *Sea Calves*.
Segonius, 321.
Seneca, 313.
Sennacherib, 209.
Seræ, *who live four hundred years*, 15.
Serpeda de Aqua, 291.
Serpents, *bite of, cured by men's saliva*, 10; *ditto by odour of men*, 11; *test of fidelity of wives*, 11; *destroy strangers*, 69; *war with Weasels and Swine*, 70; *killed by Spiders*, 71; *and Cats*, 154, 155, 156; *and Mice*, 156; *and Lions*, 156; *cure for bite of*, 161; *take medicine*, 162; *the Indian, a kind of whale*, 226, 227; *and Crabs*, 267, 268; *charming them*, 278, 279; *their loves*, 280, 281; *talking*, 281; *size*, 281, 282; *their coldness*, 283, 284; *pugnacity*, 284, 285; *their antipathies*, 285, 286, 287; *as medicine*, 288, 289.
Servius, 171.
Sextus, 134, 138.
Shrew mouse, the, 68, 70.
Shu-Maon, *a hairy man*, 49.
Sicinnis, Sicinnistæ, *a name for Satyrs*, 57.
Sidetes, 140.
Sileni, *a name for Satyrs*, 56, 57.
Simeon, Rabbi, 166, 167, 168.

Simia Satyrus, the, 52, 53, 54, 56.
Simiinæ, the, 51.
Simocatus, 286.
Sindbad the Sailor, 218.
Siren, the, 172, 173.
Sluper, John, 7, 45, 65, 229.
Snow Birds, 191, 192, 193.
Solinus, 58, 313.
Solyman, Sultan, 96.
Somerville, Sir John, 298, 299, 300.
Sow, 135, 136.
Spenser, 88, 158, 312.
Spermaceti Whale, the, 222.
Sphyngium, the, 53.
Sphynx or Sphynga, 61, 62, 63, 64, 65, 159.
Spider, the, 69, 70, 71.
Sponges, 260, 261.
Spratt, 171.
Stag, the, 68, 69, 163.
Stanley, H. M., 31, 32.
Starchaterus Thavestus, *a giant*, 33, 34, 35.
Steingo, *a name for a Gorgon*, 85.
Stheno, 85.
Sting-ray, the, 256, 257.
Stork, the, 162, 200, 201.
Stow, John, 231.
Strabo, 314.
Struthpodes, *a tribe with small feet*, 15.
Stumpsius, 308.
Su, the, 163, 164, 165.
Suidas, 65, 146.
Swallow, the, 161, 186, 187, 188, 189.
Swamfisck, the, 245, 246, 247.
Swan, the, 69, 193, 194.
Swine, 70, 148, 156.
Swordfish, the, 238, 239.
Sylla, 58.
Syrbotæ, *men twelve feet high*, 17.

INDEX.

T.

TANTALUS apples, 75.
Tauron, 15.
Tavernier, 191.
Tennent, Sir J. E., 213.
Teüfelwal, the. See *Trol Whale.*
Thenestus, 163.
Theophrastus, 106, 118, 119.
Thibii, *a tribe having two pupils to each eye*, 12.
Thos, the, 71.
Thresher-Whale, the. See *Orca.*
Tiles, *shower of baked*, 251.
Toad, the, 326, 327, 328.
Topazos, *a beautiful stone*, 21, 22.
Topsell, Edward, 53, 55, 66, 74, 83, 91, 92, 94, 97, 99, 104, 127, 131, 145, 146, 154, 163, 270, 278, 282, 288, 289, 291, 302, 306, 308, 312, 313, 317, 325, 326, 327, 331.
Tortoise, the, 161.
Traconyt, *a beautiful stone*, 21.
Tragi. See *Sponges.*
Tranquillus, 147.
Trebius, the, 252.
Trebius Niger, 254, 264, 266.
Triballi, *a tribe having the power of fascination with their eyes*, 12.
Triorchis, the, *a hawk*, 70.
Trispithami, *a race three spans high*, 27.
Trithemius, 144.
Tritons, 65, 210.
Trochilus, the, 70, 201, 202.
Troglodytæ, *dwellers in caves*, 14; *their swiftness*, 17; *their remains*, 20; *feed on serpents and lizards*, 21; *their commerce*, 22.
Trol Whale, the, 217.
Trygon, the. See *Sting-ray.*

Turtles, *horned*, 23.
Turtle-dove, the, 70.
Tytiri, *a name for Satyrs*, 56.
Tzetzes, 93.

U.

UNICORN, the, 74, 87, 88, 89, 90, 91, 92, 93, 94, 95, 96, 97. See also *Rhinoceros.*
Urchin, the, 128.

V.

VALENTYN, 213.
Varinus, 64.
Varro, 10.
Versipellis. See *Were Wolves.*
Vespasian, 151.
Vielfras, the. See *Gulo.*
Villanonanus, Arnoldus, 287.
Vipers, *flesh of, causing longevity*, 16.
Virgil, 140.
Vishnu, 209.
Volateran, 282.

W.

WALLACE, A. R., 52.
Walrus, the, 235, 236, 237, 238.
Wantley, Dragon of. See *Dragons.*
Wasp, the, 70.
Weasel, the, 68, 70, 163.
Webbe, Edward, 250.
Webber, *Romance of Natural History*, 30.
Were Wolves, 140, 141, 142, 143, 144.
Whale, the, 214, 215, 216, 217, 218, 219, 220, 221, 222, 223, 224, 225, 226, 227.
Whale, *the hairy*, 226.
Whaup, the. See *Lapwing.*

INDEX.

Whirlpool, the, 215, 220.
Williams, Edward, 189.
Woodcock, the, 69.
Wolf, the, 68, 131, 134, 135, 136, 137, 138, 139, 140, 148.
Wolff, G. E., 31.
Wolverine, the. See *Gulo*.
Wood, E. J., *book on Giants and Dwarfs*, 29.
Wood, W. Martin, 50.
"Wormes." See *Dragons*.

X.

XENOPHON, 86.

Y.

YOULE, Captain HENRY, 49.

Z.

ZAHN, JOANNES, 4, 144, 165, 173, 248.
Zaidu, 79.
Zebra, 146, 147.
Ziphius, the, 238, 239.
Zoophytes, 259, 260.

THE END.

Reprint Publishing

For People Who Go For Originals.

This book is a facsimile reprint of the original edition. The term refers to the facsimile with an original in size and design exactly matching simulation as photographic or scanned reproduction.

Facsimile editions offer us the chance to join in the library of historical, cultural and scientific history of mankind, and to rediscover.

The books of the facsimile edition may have marks, notations and other marginalia and pages with errors contained in the original volume. These traces of the past refers to the historical journey that has covered the book.

ISBN 978-3-95940-122-7

Facsimile reprint of the original edition
Copyright © 2015 Reprint Publishing
All rights reserved.

www.reprintpublishing.com

www.ingramcontent.com/pod-product-compliance
Lightning Source LLC
Chambersburg PA
CBHW050154230526
45470CB00001B/83